RESEARCH METHODOLOGY
FOR ENGINEERS

RESEARCH METHODOLOGY
FOR ENGINEERS

R. Ganesan

Advisor (Academic Research)
and
Professor of Eminence
Department of Mechanical Engineering
B.S. Abdur Rahman University, Chennai

MJP Publishers

    No. 44, Nallathambi Street,
Triplicane, Chennai 600 005

MJP 098   © Publishers, 2024

Publisher : C. Janarthanan

PREFACE

Research Methodology has been a subject in the curriculum of Business Management and Master of Philosophy programmes of many universities. These programmes belong to the humanities and social science discipline. Recently, the University Grants Commission (UGC) made it mandatory that Research Methodology should be included as a compulsory course for all Ph.D. and M.Phil. programmes offered by all Universities in India. While many books have been written to meet the requirements of research scholars of humanities and social sciences, not many books addressed to meet the needs of scholars pursuing research in the field of engineering and physical sciences are available.

Researchers in the field of humanities and social sciences often employ survey methods either of the physical or questionnaire type or in-depth case studies as a part of their research method. But, researchers in the field of engineering or physical science resorts to experimental methods and/or simulation approaches as a part of their work. While the books available on Research Methodology are heavily oriented towards survey methods, etc., not many books on Research Methodology dealing with experimental and/or simulation approaches are available. However, a reader can see some excellent books on the subject of experimental methods and use of statistical tools for analysing experimental results available.

It is felt that books which exclusively cover the needs of research scholars holistically in the field of engineering and physical science are not many. It was this aspect which provided the necessary motivation to venture into writing this book. I have tried to provide a holistic treatment to the subject of Research Methodology, which I believe that a research scholar pursuing research in the sphere of engineering and science should be conversant with. Therefore, the targeted audiences are predominantly academic research scholars of engineering and science.

This book is organized into eighteen chapters. Chapter 1 deals with certain basic concepts of research, research process and attributes of good research scholar. How to formulate and define research problem forms the subject matter of Chapter 2. Concepts of research design and some important experimental designs are covered in Chapter 3. Development of mathematical models and analysis is often resorted to by engineers as a part of their research work. Chapter 4 deals with mathematical modelling and validation giving case examples. Simulation is another important tool frequently used by engineers and scientists to analyse situations. Similarly dimensional analysis is yet another tool often employed by engineers. Chapter 5 covers both these aspects.

Statistical technique finds extensive use in engineering research studies. An attempt has been made to cover important statistical concepts needed by a research scholar. While Chapter 6 deals with probability and distributions, Chapter 7 is devoted to sample design and sampling, and Chapter 8 covers hypothesis testing and analysis of variance. Design of experiments and regression and correlation analysis forms the subject matter of Chapter 9. As statistical method is a complete subject by itself, full coverage is beyond the scope of this book. Only significant aspects relevant to the research scholar in engineering and science are included. In addition, use of Taguchi's orthogonal array in designing experiments economically has been dealt with.

Analysis and interpretation of data is a crucial step in a research study. Realizing this, Chapter 10 discusses this aspect. Many researchers do not give due recognition to error analysis of the results of their experiments. Repeatability, reproducibility and error analysis are treated in Chapter 11. There is good scope for use of optimization techniques in experimental studies. Many researchers currently use evolutionary optimization techniques in their research studies. Hence, an introduction to optimization techniques is given in Chapter 12.

A research scholar apart from being a good researcher, should develop adequate skills and capabilities in preparing papers for publication in refereed journals and write synopsis of his/her research work. Chapter 13 is included to meet this need. Culmination of research work is the preparation of thesis, a standalone document for evaluation by examiners. Chapter 14 and 15 are exclusively devoted to cover preparation of thesis. A suggested structure of thesis is also given in Chapter 15.

There are no universal guidelines for evaluation of thesis unlike regular academic programmes by study. It is dependent on the examiner's perception. Chapter 16 concentrates on practices being followed in evaluation. A researcher should confidently face the oral examination and defend his/her thesis. This aspect is addressed in this chapter. Ethics in research is an important issue and hence ethics and integrity in research is covered in Chapter 17. Intellectual Property Rights (IPR) is yet another important issue, a researcher should be aware of in the present competitive scenario. Recognizing this, Chapter 18 covers aspects like Copyrights, Patent law, etc. especially relating to Indian situation for the benefit of the research scholars.

The reader may find many of the concepts and theory covered in this book available in many published literature. I have endeavoured to provide relevant concepts and methods and reflect it in a form that may be required by academic research scholars of the engineering discipline for their work. My experience in teaching this subject to research scholars in the different engineering disciplines has helped me in this endeavour. This book would definitely satisfy the felt needs of such scholars.

The merit of this book, I believe, lies in the wholesome treatment of Research Methodology for the benefit of research scholars. As not many books are available to meet the needs of the targeted audience, this book would serve as a useful resource material.

R. Ganesan

ACKNOWLEDGEMENTS

As I was toiling with the idea of preparing a book on research methodology for engineers, the spontaneous support and encouragement provided by Professor Dr. V.M. Periasamy, Registrar of B.S. Abdur Rahman University, Professor Dr. M. Abdullah Khan of Electrical and Electronics Engineering Department and Professor Dr. V. Sankaranarayanan of Information Technology Department gave the necessary courage and impetus to undertake this venture. I would like to place on record my heartfelt thanks to all of them for their support and encouragement. Professor Dr. R. Srinivasa Raghavan of Civil Engineering Department made several suggestions while preparing the contents and his help is greatly acknowledged.

The research work of Dr. M. Murugan, Dr. K.M. Abubacker, Ms. Nasreen Kaleem of Mechanical Engineering Department, Dr. R. RajaPrabu of Electrical and Electronics Engineering Department and Ms. Sabura Banu of Electronics and Instrumentation Department have been utilized to prepare case studies and examples. I am greatly indebted to them for their gesture. Dr. Raghavachari Rajan of Unilever was kind enough to go through the first draft of this book and make several comments and suggestions for improvement and I thank him and place on record hisvaluable contribution and spontaneous help.

Special recognitions and thanks are due to my personal secretaries Ms. Jabeen Rehana and Ms. C. Jeyanthi for their unstinted efforts and hard work in transcribing my manuscript into the electronic media in a neat and presentable fashion. Mr. R. Tamil Selvan, Draftsman of Mechanical Engineering Department has taken pains to prepare the sketches for the book and my special thanks are due to him. Finally, I wish to express my sincere gratitude to all others who have helped me in one way or other in bringing out this book.

I gratefully acknowledge all the source books from where I got the basic inputs to write this book on research methodology. I express my sincere thanks to the authors of these valuable books.

Last but not the least, I wish to thank Mr. J.C. Pillai, Director, and Mr. C. Sajeesh Kumar, Managing Editor of MJP Publishers, and their team for the excellent editorial work and for bringing out this book in its present form.

R. Ganesan

CONTENTS

1
Research Process

1.1 RESEARCH

1.1.1 Significance of Research

> A knowledge economy is one that relies extensively on human skills and creativity, the utilization of human intellectual capital supported by lifelong learning and adaptation, the creative exploitation of existing knowledge, and extensive creation of new knowledge through research and development.
>
> **Anonymous**

The quality of life of a citizen of a country greatly depends on the overall development the country achieves. We live in an era characterized by rapid changes, uncertainty and competition where knowledge is an indispensable ingredient for sustained growth and development. New knowledge has to be created and embodied in technologies, products, processes and services. We have witnessed that the economic development of many countries is on account of sustained and substantial investment in research and development activities both by the state and the private sectors among other measures.

While the nineteenth century was the century of the Europeans, the twentieth century became the century of the Americans due to their leadership in the technological forefront and concomitant economic power that took them to the top. The twenty-first century can be that of Asians. While Japan showed its prowess in the last quarter of the twentieth century followed by South Korea, Singapore, etc., the twenty-first century will focus more on India and China.

Research is needed in all facets of economy such as governmental policies, planning and operation of industry and business, social sciences, agriculture, health, etc. for the progress of the society. Whatever may be the field, the contribution of research through science and technology is highly significant. In fact science and technology are driven by the society and, they are interrelated. Research and development are an essential feature of any developed economy. For the economic supremacy of the USA, the contribution of science and technology through research is a prime factor. Realizing the importance of research and development, China, which is a fast growing economy declared that 60% of GDP will be related to science and technology in the next two decades.

We as Indians have demonstrated that we are capable of absorbing any new technology, but we have not improved our track record with regard to innovation of new technologies. There is a gap in the technological competence. This gap can be eliminated by creative talents. Such talents in the last two decades were attracted by the western world from India. During the last two decades, our computer specialists have made significant contributions outside the country. We have to make a systematic approach to contribute more in the technological front, be it engineering, medicine, agriculture or other fields, within our country.

It is a matter of concern to note that while USA has 4700 researchers per million population, India has a very low figure of 156 per million population. In absolute numbers, the USA has 1.57 million researchers, China with 1.42 million, India accounts for only 0.154 million. India as a developing economy has to pay more attention to research and development in science and technology. Inquiry or inquisitiveness is the basic tenet of research. As per Hudson's maxim, "All progress is born of inquiry. Doubt is often better than overconfidence, for it leads to inquiry, and inquiry leads to innovation." Increased amounts of research ensure progress.

Increasing global competition and disbanding of trade barriers drive many countries to commit more resources for research and development activities both in industry and academia. Our Central Government has recognized this and in its eleventh five-year plan has increased the allocation in the budget for R&D and higher education. Apart from this, efforts are being made to revamp higher education, creation of higher centres of learning and universities and upgradation of reputed educational institutions to the status of university. The University Grants Commission places more emphasis on research. There is an increasing awareness of the importance of involving industries with academia to undertake applied research to address the problems of industry and society.

1.1.2 Research: What it connotes

According to the Oxford Dictionary, the word "research" connotes the systematic investigation into and study of materials, sources, etc., in order to establish facts and reach new conclusions. It also indicates "an endeavour to discover new or collate old facts, etc., by the scientific study of a subject or by a course of critical investigation".

The word "research" may mean different things to different people. It has been defined in several ways. Nevertheless we should have a clear picture as what is relevant to us. It is a scholarly or scientific investigation or inquiry and it contributes to the expansion of knowledge through discovery of facts, theories or laws. It may involve hunting for facts or experimentation.

Through research, we realize the goal of science, thus expanding the frontiers of knowledge. This requires the process of organized reflective contemplation by the researchers.

Reflective contemplation is nothing but a planned and serious thinking about an aspect which may ultimately result in an innovative idea or a new concept, evolving a solution to a problem. The development in science and technology is on account of such organized knowledge creation. This may involve planned observation, identification, description, experimentation or theoretical explanation of any phenomenon. Thus, research promotes development in science and technology, which further widens the knowledge horizon.

As far as industry research and development is concerned, it refers to the work directed towards the innovation, introduction and improvement of products and processes. It is a voyage of discovery propelled by the instinct of inquisitiveness of the human mind. It is an academic activity and as such the term should be used in the technical sense. According to Clifford Woody, research comprises defining and redefining problems, formulating suitable solutions or hypothesis; collecting, organizing and evaluating data; making deduction and reaching conclusions; and at last carefully testing the conclusions to decide whether they satisfy the hypothesis formulated.

Research is not confined to science and technology alone and it pervades all disciplines such as history, language, sociology, etc. Whatever be the discipline, it has to be an active, diligent and systematic process of inquiry in order to discover, interpret or revise facts, events, behaviour or theories. Cross fertilization of ideas from different disciplines also falls in the realm of research and development.

There are several examples wherein cross breeding of ideas from one discipline to another have yielded significant results. Many biological methods and systems found in nature have been extensively used in the design of engineering systems. For instance, the design of Velcro by the Swiss engineer George de Mestrol is based on his observation as how the hooks of burrs clung to the bur on his dog. In computer science cybernetics tries to model the feedback and control mechanism taking clue from the inherent and intelligent behaviour of biological systems. Similarly in the area of optimization, evolution of optimization techniques such as Particle Swarm Optimization (PSO) have been evolved due to the study of organizational principles from the social behaviour of organisms such as the flocking behaviour of birds. In material development certain paints and roof tiles have been engineered to be self-cleaning types, copying the mechanism from the Nelumbo lotus.

Research is an original contribution to the existing stock of knowledge making for its advancement. We try to establish the truth with the help of study, observation, comparison and experiment. In short, the search for knowledge through an objective and the systematic way of finding solution to a problem is research.

1.1.3 Objectives and Motivation

The primary objective of research is to find answers to questions through the application of scientific procedures. The truth is hidden and has not been found out so far. Every research has its own specific purpose, but one can broadly group the research objectives as follows:

- To gain familiarity with the phenomenon or to get new insights into it. These studies are termed as exploratory or formative research studies.

- To determine accurately the characteristics of a particular individual, system, situation or a group. These studies are known as descriptive research studies.

- To decide the frequency with which a phenomenon occurs or with which it is associated with something else. These studies are known as diagnostic research studies.

- To establish the causal relationship between variables by forming suitable hypotheses. They are termed as hypothesis-testing research studies.

There are several motives for somebody undertaking a research work. The possible ones include the following.

- Desire to get a research degree along with its consequential benefits.
- Desire to face challenges in solving the unsolved problems.
- To get intellectual joy of doing some creative work.
- To do service to the society and gain respect.
- Direction by the employer, government, etc.

Research should not be viewed as a mundane activity, the sole purpose being to secure a doctoral qualification to satisfy some administrative or promotional needs. The scholar should be motivated by the spirit of inquiry, genuine interest of making useful contribution to the literature in the field and to solve the problem of the society. In fact, the scholar should view the research in a noble sense.

1.2 THE RESEARCHER

1.2.1 Creativity and Problem-Solving

Creativeness is an essential trait for a good researcher. Creative thinkers are distinguished by their ability to synthesize new combinations of ideas and concepts into meaningful and useful forms. According to Brian Tracy, one of the top management consultants, "creativity is a natural and normal ability, possessed in quantity by virtually everyone. It is inborn, a part of your genetic structure, a faculty that is uniquely human. Everyone is creative". He adds that "creativity is like a muscle. If you don't use it, you lose it. Just like a muscle, if you do not exercise your creativity and stretch it regularly, it becomes weak and ineffective. Your ability to generate ideas must be constantly utilized to be kept in top condition".

Creative process can be viewed as moving from an amorphous idea to a well-structured idea, from the chaotic to the organized, from the implicit to the explicit. As Brian Tracy observed, creativity can be cultivated and entranced with study and practice. Some of the positive steps to enhance creative thinking is given below.

- Develop a creative attitude. To be creative, it is essential to develop confidence that one can provide a creative solution to the problem. Confidence comes with success, so start small and build confidence up with small success.

- A researcher should unlock his/her imagination. He/she should rekindle the vivid imagination he/she had as a child.

- Be persistent. Creativity requires hard work. Most problems will not succumb to the first attack. They must be pursued with persistence. Edison tested over 6000 materials before he discovered the species of bamboo that acted as a successful filament for the incandescent light bulb. It was also Edison who made the famous comment, "Invention is 95 percent perspiration and 5 percent inspiration".

- Develop an open mind. This implies receptiveness to ideas from any and all sources. Many a times cross-breeding ideas from different disciplines help to solve a problem.

- Hasty and earlier judgment must be avoided since creative ideas develop slowly.

- Establishing the boundaries of a problem is an essential part of problem definition. In fact it does not limit creativity, but rather focuses it.

- One should cultivate vertical or convergent thinking as well as horizontal or lateral thinking.

1.2.2 Attributes of a Good Research Scholar

A research scholar should be motivated by the spirit of achievement and contribution. The attributes or trails of a good research scholar could be the following.

- should be honest, flexible and self-confident.
- should possess analytical mind, insight, intelligence and innovative approach.
- should be free from obsessions of time and willing to work long hours.
- must have intellectual curiosity, optimism, patience and persistence.
- should be willing to work with dedication and determination.
- should be a keen observer and have spirit of enquiry and laboratory temperament.
- should have the necessary resilience to withstand temporary setbacks.
- should have passion for knowledge, scientific temper and originality.
- should have good communication skills, both oral and written.
- should possess organizational ability and be willing to work in a team.

1.3 TECHNOLOGICAL INNOVATION

Engineering is concerned with problems whose solution is needed and/or desired by the society. The capacity to innovate, manage information, and nourish knowledge as a resource will dominate the economic domain as natural resources, capital and labour once did. This places high premium on development of technology and delivery systems.

The advancement of technology has three phases as given below:

Invention   The creative act whereby an idea is conceived.

Innovation The process by which an invention or idea is translated into successful practice and is utilized by the economy.

Diffusion   The successive and widespread initiation of successful innovation.

Science-based innovation has made significant contributions to the development of industries like aircraft, computers, plastics and television in developed countries like USA. Traditionally, engineers play a major role in technological innovation. A strong basic research is needed to maintain the storehouse of new knowledge and ideas. Innovation in response to a need of the society has greater probability of success than innovation in response to technological research opportunity. The scope for doing applied research is high, which involves innovative solution to the problem of the society.

1.4 TYPES OF RESEARCH

Research approaches are classified on the basis of one or more of the following.

- whether the study involves any exploration of the existing system
- whether any fundamental issues are involved
- whether the analyst uses quantitative or qualitative information or whether it involves physical experiments or virtual experiments using computer.

The basic types of research are as follows.

1.4.1 Descriptive vs Analytical

The major purpose of descriptive research is description of the state of affairs as it exists at present. This method includes surveys and fact-finding enquiries of different kinds. The main characteristic of this method is that researchers have no control over the variables, they can report only what is happening or what has happened (e.g., frequency of arrival of vehicles for repairs in a workshop). In analytical research, the researcher has to use facts or information already available from secondary sources and analyses them critically so as to derive conclusions.

1.4.2 Applied vs Fundamental

Applied research is also known as "action research". It aims at finding a solution for an immediate problem facing the society or an industrial/business organization. On the other hand, fundamental research is concerned with generalization and with the formulation of a theory. Research concerning some natural phenomena or relating to pure mathematics are examples of fundamental research. The central aim of applied research is to discover a solution to some pressing practical problems, whereas basic research is directed towards finding information that has a broad base of applications and thus adds to the already existing organized body of scientific knowledge.

1.4.3 Quantitative vs Qualitative

Quantitative research is based on the measurement of quantity or amount. This is applicable to phenomena that can be expressed in terms of quantity, amount, size or weight, etc. Qualitative research on the other hand is concerned with the qualitative phenomena, that is, something relating to or involving quality. For example, research designed to find out how people feel or what they think about a particular subject or institution is qualitative research.

1.4.4 Conceptual vs Empirical

A concept is an abstract idea, a visionary expression of a proposed or planned action that leads to achievement of an objective. Therefore, a conceptual research is related to some abstract idea or theory. It is generally used by thinkers to develop new concepts or reinterpret existing ones. On the other hand, empirical research relies on experience or observation alone, often without due regard for system and theory. It is data-based research ending up with conclusions, which are capable of being verified by observation or experiment.

Many engineering-oriented research work are of empirical type involving physical experiments or simulation. In such a research, the researcher tries to establish a working hypothesis (probable) and then tries to get enough facts (data) to prove or disprove his/her hypothesis. The researcher then sets up experimental design to manipulate the material concerned so as to bring forth the desired information. In this type of research, the experimenter has control over the variables under study and can deliberately manipulate any one of them to study its effects. Evidence gathered through experiments or empirical studies is considered to be the most powerful support for a given hypothesis.

1.4.5 Other Types of Research

Other types of research are mostly variations of one or more of the above-stated approaches and include,

One-time or longitudinal   One-time is confined to a single period whereas longitudinal is carried over several time periods.

One-time or cross-sectional study envisages the study of a population by drawing samples from that population, and perform critical examination. Here, we do not manipulate the variables; instead, we try to examine their possible correlation, association, etc. This type of study does not adduce any evidence to conclude that a particular variable is the root cause of an event of interest.

The longitudinal research study, on the other hand, is an experimental investigation. In the longitudinal research, the researcher manipulates the input variables and factors either using physical experiments or through virtual simulation with the aid of ICT tools and make observation of the output variable(s) of interest. This type of research studies provide us tangible evidence to determine the input variable(s) that had really contributed to the performance of the output variable.

Field-setting, laboratory or simulation   This is based on the environment in which the research work is carried out. For instance a research study is made to identify the factors responsible for the low productivity of a manufacturing organization. In that case, the factory provides the field-setting. Physical experiments are designed to meet the scope of a particular study and the laboratory provides field-setting to conduct experiments under controlled conditions. There are many situations wherein experiments and field study are neither possible nor economical to conduct. In such situations a model is developed and simulation study is conducted using a computer.

Conclusion-oriented and decision-oriented   In conclusion-oriented research, a researcher is free to pick up a problem, redesign the enquiry as he proceeds and is prepared to conceptualize as he wishes. Decision-oriented research is always to meet the needs of the decision-maker, and the researcher is not free to embark upon research according to his/her own inclination. Operations research is an example of decision-oriented research.

From the above description of the types of research, it is obvious that there are two basic approaches to research, viz., quantitative approach and qualitative approach.

As already stated the quantitative approach involves generation of data in quantitative form using cardinal numbers, which are subjected to rigorous quantitative analysis in a formal and rigid fashion. This approach can be further subdivided into (i) inferential (ii) experimental (iii) simulation approaches. In the inferential approach, the characteristics of the population are inferred from the database. The variables are manipulated by the researcher in the experimental approach to observe the effect of a variable on others. Greater control is exercised over the experiment. The simulation approach involves building a numerical model and manipulating the model with the help of a computer. This permits observation of the dynamic behaviour of the system or subsystem under controlled conditions.

Qualitative approach to research is concerned with subjective assessment of attitudes, opinions and behaviour. This approach generally generates results in non-qualitative form or in a form, which is not subjected to rigorous quantitative analysis.

1.5 RESEARCH VS SCIENTIFIC METHOD

These two are closely related. Research can be termed as "inquiry into the nature of reasons and the consequences of any particular set of circumstances, whether these circumstances are experimentally controlled or recorded just as they occur. The researcher is interested in more than one particular results; he is interested in the repeatability of the results and in their extension to more complicated general situations."

The term "scientific method" covers the philosophy common to all research methods and techniques, although they may vary considerably from one science to another. The scientific method is one and the same in the branches of a particular science. This method encourages a rigorous, impersonal mode of procedure dictated by the demands of logic and objective procedure.

In scientific method, the scientist with the existing background of knowledge propelled by curiosity starts questioning the laws of science. As a result of such questioning he/she formulates a hypothesis, which is subjected to logical analysis and the hypothesis is either accepted or rejected. When the analysis does not lead to any tangible results either due to flaws or inconsistencies, the scientist modifies the hypothesis and repeats the analysis. Once the result of the study confirms the hypothesis, it is accepted as a proof by other scientists and is added as a contribution to the existing knowledge base. Thus, the knowledge base gets enlarged.

As already stated, research differs from scientific method only in viewpoint and philosophy. Here also, literature review leads to the identification of the state-of-the-art of knowledge in the aspect considered by the researcher. The aspect the researcher has taken into consideration may include scientific knowledge, devices, components, materials, processes, economic and social problems such as environment, etc. As far as academic research is concerned, the aspiration or inquisitiveness of the researcher to contribute to the literature so as to satisfy his/her objective of undertaking such research work provides the impetus. Academicians also do undertake applied research, which the industry develops further before commercialization of the product or processes evolved as research outcome.

However, in the industrial environment the need for a particular research is mostly driven by the needs of the customer based on input from the marketing personnel, or the firm may want to reduce the cost by evolving different processes or by employing a newer material or due to restrictions imposed by governmental agencies. For example, emission norms fixed by Governments compelled the automotive industry to undertake several research projects involving improving of fuel performance, modification of engines, engine management system through microprocessors, etc. Such type of research work is more relevant to meet the goals of the organization.

1.6 RESEARCH METHODOLOGY VS RESEARCH METHODS

Research methods refer to the methods the researchers use in performing research operations. For instance, in applied research, the objective is to arrive at a solution for a given problem;

the available data and the unknown aspects of the problem have to be related to each other to make a solution possible. For this, the research methods employed can be grouped into i) methods used to collect data ii) statistical techniques used for establishing relationships between data and unknowns, and iii) methods used to evaluate the accuracy of the results.

The scope of research methodology is wider than that of the research methods. Here, we not only talk of research methods, but also consider the logic behind the methods we use in the context of our research study and also explain why we are using a particular method or technique and why we are not using others. This is necessary to ensure that research results are capable of being evaluated either by the researcher or by others.

The researcher needs to know which method or technique is relevant and which is not, what they would mean and explain why. The researcher also needs to understand the assumptions underlying various techniques and know the criteria by which they can decide that certain techniques and procedures will be applicable to certain problems and others will not. All this mean that, the researchers have to design their methodology for their problems as the same may differ from problem to problem.

A researcher working in the field of engineering may resort to collection of all background data relevant to his/her study from secondary sources such as published literature, handbooks, etc., while a researcher working in the field of social science such as marketing may resort to gathering data from competitor's literature, trade publications, etc. With regard to primary data, the engineering researcher may employ experimental methods or virtual simulation methods wherein personal observation of output can be made with the help of measuring instruments, while a social scientist may resort to questionnaire survey.

To establish relationships between data and unknown, both engineering and social science researchers may adopt statistical techniques like correlation and regression analysis or mathematical techniques. To ensure creditability of the methods used and the results obtained, the researcher validates the methodology adopted by comparing the result with past performance and also carries out error analysis using appropriate statistical methods to evaluate accuracy of results.

Suppose a researcher has used linear programming method for analysis for a particular research study with data collected, it is incumbent on his/her part to justify as to how the linear programming is a suitable OR technique in that particular context. Perhaps the researcher may have to justify on the grounds of linearity of relationship between variables and the additive contribution they make to the objective of interest.

1.7 RESEARCH PROCESS

The research process consists of series of actions or steps necessary to effectively carry out research and the desired sequencing of steps. Figure 1.1 illustrates the research process.

It may be seen that the research process consists of several closely relatd activities, which overlap considerably

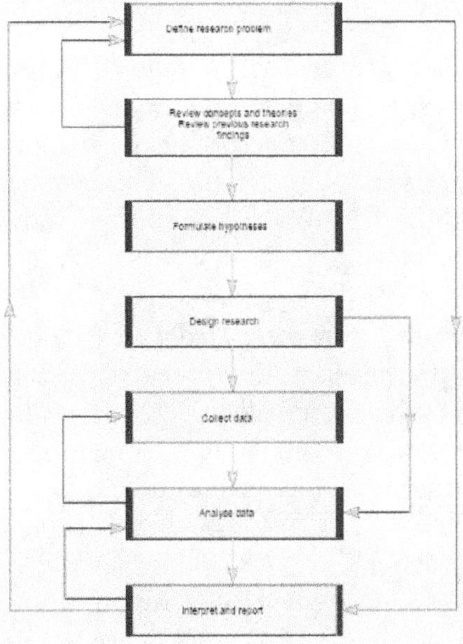

Figure 1.1   Flowchart of the research process

rather than strictly following a prescribed sequence. Many a times, the first step determines the nature of the last step to be undertaken. It is better that subsequent procedures are taken into account in the early stages, to avoid serious difficulties later. The various steps involved in the research process are not mutually exclusive, nor are they separate and distinct. They do not necessarily follow each other in any specific order and the researcher has to constantly anticipate at each step the requirements of subsequent steps.

The following sequence of steps provides a useful procedural guideline regarding the research process.

   i.   Formulating the research problem

   ii.   Extensive literature survey

   iii.   Developing the hypothesis

   iv.   Preparing the research design

   v.   Determining sample design

   vi.   Collection of data

   vii.   Project execution

   viii. Data analysis

   ix.   Testing the hypothesis

   x.   Interpretation and generalization

   xi.   Report preparation

Formulating the research problem   At the outset the researchers must decide the general area of interest or aspect of subject matter that they would like to inquire into. The formulation of a general topic into a specific research problem constitutes the first step in a scientific enquiry. Essentially two steps are involved in formulating the research problem, viz., understanding the problem thoroughly, and rephrasing the same into meaningful terms from an analytical point of view.

Literature survey   Literature survey connected with the problem is very essential for a Ph.D. research work. One of the major reasons is that the researcher should get acquainted with the selected problem both conceptually, i.e., concepts and theories, and empirically by reviewing studies made which are similar to the one proposed. This gives confidence in defining the problem and also in identifying and narrowing the topic of research.

Development of working hypothesis   A working hypothesis is a tentative assumption made in order to draw out and test its logical or empirical consequences. In most of the research work, it plays an important role. The role of hypothesis is to guide the researchers by delimiting the area of research work.

Research design   The function of a research design is to plan for the collection of relevant data/evidence with minimum expenditure of effort, time and money. But this depends mainly on the research purpose. Research purposes may be grouped into

   i.   Exploration,

   ii.   Description,

   iii.   Diagnosis, and

   iv.   Experimentation.

Sample design   The field of inquiry cannot constitute the entire "universe" or population as it involves great deal of time, money and energy and many times may be impossible. Hence, quite often we select a few items from the universe, i.e., a sample, for our study purpose. The way a researcher decides to adopt in selecting a sample is popularly known as "sample design". There are several ways of selecting a sample—random sampling, systematic, stratified, sequential, multistage, cluster, etc.

Data collection   Primary data required for the research study can be collected either through experiment or through survey. If the researcher conducts an experiment, he/she observes some quantitative measurements or the data, with the help of which the truth contained in the hypothesis is examined. The secondary data is gathered from published reports, handbooks and literature.

Project execution　This is an important step in the research process. The execution of the entire project should have been planned properly. If the execution proceeds on correct lines, the data collected would be adequate and dependable. Many a times, it will be useful if a pilot experiment or preliminary survey is conducted before proceeding to the actual production run.

Data analysis　The analysis of data requires a number of closely related operations such as establishment of categories, the application of these categories to raw data through coding, tabulation and then drawing statistical inferences. If the data is voluminous, use of computers and statistical packages can be resorted to.

Testing of hypothesis　In this step, the hypothesis formed already is tested using various statistical tests such as Chi-square, T-test, F-test, etc. The hypothesis testing will result in either accepting the hypothesis or rejecting it.

Interpretation and generalization　If the researcher had no hypothesis to start with, the findings may be explained on the basis of some theory. It is known as interpretation. Quite often the process of interpretation may lead to new questions, which in turn may lead to further researches. If a hypothesis is tested and upheld several times, it may be possible for the researchers to arrive at generalization that is to build a theory. The real value of research lies in its ability to arrive at certain generalization.

Reporting or thesis writing　Reporting or thesis writing is the culmination of the research work done. Generally it is a complete, stand-alone document aimed at persons having good background of knowledge especially in the case of a Ph.D. thesis. Enough time should be devoted for writing the thesis and it should be in proper format. Manuals are available to guide the researcher. It is always better for the researcher to go through some of the well-written theses to grasp the nuances.

1.8 RELEVANT AND QUALITY RESEARCH

Relevant and quality research is essential for development, which in turn contributes to the development of national wealth, improve the quality of life and enhance national security. Both basic research and applied research are necessary especially in the current competitive global scenario. While basic research needs the support of better technology-enabled equipment and facilities, applied research needs proper interaction of universities with and patronage of industries, Government and other related organizations. In the space science and atomic energy-front we try to achieve self-sufficiency to a greater extent in meeting the basic research needs. Nevertheless, we have to modernize our basic research in other aspects undertaking research in cutting-edge areas that enable us to globally remain competitive and provide input to applied research.

Barring well-established institutions like IIScs, IITs and national defence and space laboratories, many universities are not adequately equipped for pursuing basic research in cutting-edge areas. With the growing awareness both on the part of Government and the academia, it is felt that the situation would improve in the future.

As far as applied research is concerned, it could be mission-oriented for technology development and problems related to industries. For the development of applied research, the involvement of industries with academia is very essential. There has to be a marked change in lateral thinking between the two, rather than vertical thinking by each. One of the major reasons why academia desists from doing applied research is lack of appreciation and support by industries technically and financially. The trend has to change.

The main motivation behind many faculty members undertaking research in the present day is to meet the requirement for promotion or to add on an additional qualification. This mindset has to change and the primary motivation should be for contribution to the development of science and technology, thereby to the society of whom they are a part. The spirit of contribution is a sine-quo-non not only for producing quality in research but also for undertaking studies in cutting-edge areas.

It is often commented that we do right research at the wrong time. This is in the context of relevance of research in the modern era. The other comment is that many of our research works are lacking in quality as is evident from the number of publication of articles in top refereed journals of international repute. Research articles contributed by Indian authors in international journals is less than one per cent.

Research work is an intellectual work and requires commitment, dedication and an inquisitive mind. It is therefore necessary that our research scholars should take interest to work in the cutting-edge areas and to produce quality work. If there is a will there is a way. What is needed is a proper mindset. In order to choose a relevant research topic/ area, the researcher/scholar must regularly update his/her knowledge in the niche area chosen by referring to reputed journals, abstracts, reports, proceedings of conferences and seminars and other related materials besides browsing the websites of reputed Universities to know about frontier areas of researches pursued. To ensure better quality, not only is the research scholar responsible, but also equally, arethe supervisor and the evaluators of the research work.

2
Research Problem formulation

2.1 PROBLEM-SOLVING IN ENGINEERING

While the research approach presented in Chapter 1 provides a general framework, there are two fundamental approaches to solving engineering problems that arise in the discovery of knowledge, and its application to society's needs. They are

    i.   Theoretical (Physical/Mathematical) modelling

    ii.   Experimental measurement

In engineering, this is true regardless of the discipline such as mechanical, electrical, etc. or the engineering function such as design, development, research, manufacturing, etc. While some problems are adequately treated by using only theory or only experiment, most require a judiciously chosen mix of these techniques. Even though each specific application has its own peculiarities, there are some important general characteristics of the theoretical and experimental methods which will be helpful in deciding the proper blend when a choice is necessary. This will help us in organizing our thinking about the whole process.

2.1.1 Comparison Between Theoretical and Experimental Process

i.   All physical principles and their mathematical expressions (mathematical models) are approximations of the real world behaviour. Though by making the model more complex, the quality of approximations can be improved, perfection is not always possible. Practical engineering always suffers from constraints of time, money, facility, etc. Therefore depending upon the overall project objective, a researcher can choose an experimental method, which is simpler and less accurate and may be judged as adequate for the purpose.

    It may be noted that this comparison centres on the fact that theories are always approximations involving simplifying assumptions, whereas experiments when properly designed and executed or run on the actual system reveal true behaviour. This comparison favours experiment over theory.

ii.   The second comparison is based on the generality of results. The main reason for preferring theoretical approaches (when their accuracy is adequate) is that theoretical results are usually applicable to a whole class of problems whereas experimental results are peculiar to the specific apparatus on which the measurements were made. The ability to solve entire classes of problems with a single analysis gives theoretical methods an efficiency which is their major advantage. A significant price paid for the generality of theory is the universal need for simplifying assumptions.

However, when models are made more closer to the real world, it leads to mathematical formulations of deeper complexity and mathematical intractability. Today, of course, this problem has been greatly alleviated by the development of computerized numerical methods (e.g., FEA). These methods sacrifice generality and they do not provide closed solutions, rather they give only numerical answers for individual specific cases, analogues to physical experiment.

Just as pursuit of accuracy in mathematical modelling leads to undesired complexity, the experimental approaches also generally become more intricate when we require higher accuracy in the measured result. Thus, this comparison shows no obvious or uniform advantage for each approach.

iii. The third comparison can be based on facilities needed to undertake theoretical or experimental studies. Theoretical studies may be initiated with less commitment of financial resources, while experimentation requires investment in equipment, laboratory space, etc. This obvious advantage partially disappears when we realize that most current theoretical studies require good computing facility supported by suitable software. Therefore, neither approach has clear superiority.

iv. The next comparison is based on the time required to achieve results. Any technical endeavour has an appropriate timescale associated with its success. Deciding the time frame is difficult at the beginning of the project. Theory seems to offer advantages, since one can start immediately, whereas experimentation requires the building of experimental set-up, debugging, pilot run, calibration before the first data is available. However, if the chosen theoretical approach turns out futile or ends in a fiasco, we may have to reformulate, which again requires more time.

It should be clear that sometimes, a theoretical approach is quickest and at other times experimentation gives the most rapid solution. In general, simple problems are more likely to yield quickly to theory, whereas more complex problems may require a strong component of experiment.

2.2 IDENTIFICATION OF RESEARCH TOPIC

2.2.1 Selecting the Problem

In the research process, the first and foremost step happens to be that of selecting and properly defining a research problem. A researcher must find the problem and formulate it so that it becomes susceptible to research. The research problem undertaken for study should be carefully selected. The task is difficult, though it may not appear to be so. The guide may be of some help to the researcher in this. Nevertheless, every researcher must find out his/her own salvation, as research problems cannot be borrowed. A problem must spring from the researcher's mind. For instance while prescribing glasses for eyes, the optician alone cannot decide, we have to cooperate with him to get the right type of glass prescribed. A research guide can at the most only help a researcher choose a subject.

The following points may be taken into account by a researcher while selecting a research problem or a subject for research.

- Subject which is overdone should not be normally chosen, as it will be difficult to throw any new light in such a case.
- A controversial subject should not become the choice of an average researcher.
- Avoid too narrow or too vague problems.
- Choose a familiar and feasible subject so that the related research material or sources of research are within one's reach. An expert or supervisor already engaged in similar research work or journal publications will be of use in this.
- Importance of the subject, the background of the researcher, costs and time factor involved are a few other aspects which should be considered.
- Selection of a problem must be preceded by a preliminary study if needed.

2.2.2 What Constitutes Research

In a research work leading to the award of a Ph.D. degree, the researcher always looks for some original contribution. The originality under normal circumstances may be confined to,

- New interpretation or application of known facts.
- Bringing new evidences to support a known fact.
- Empirical work of one's choice and not a repetition of what someone has already done.
- Using a known technique in an untested area.
- Furthering knowledge by continuing the work suggested by another researcher at the end of the latter's research work.
- Propounding a new concept or a formula in the chosen discipline or subject.
- Evolving a new methodology to solve an existing problem.

Many scientific facts and the areas in which they have been applied to solve problems and/or used for expository analysis of situations are known clearly from literature. The researcher who is trying to identify a problem for research may thoroughly analyse these facts, develop sufficient expertise in them and then using this knowledge of the facts, can try to extend the concept to a newer area, hitherto not tried out, and successfully implement it. For example, many researchers have applied artificial intelligence tools such as Genetic algorithm to solve optimization problems in different areas as a part of their research work. The researcher may also use the scientific facts which are thoroughly familiar to him/her to explain new phenomena in a logical way not explained so far convincingly by anybody in the reported literature.

A researcher has solved a problem in a novel way and thereby claims originality. Nothing prevents some other researcher from attempting to solve the same problem which has been attempted by nobody and in an altogether different way. Especially in math-related fields, in the development of new algorithms for solving practical or hypothetical problems, this approach is adopted.

It is quite likely that while reviewing literature in a chosen field, a researcher may get a bright idea of doing an experiment altogether in a different way. The researcher may also find gaps in the assumptions of the previous researcher and the limitations in that work. Taking all these into account, he/she may evolve a better empirical method and also take care of the limitations and improve the results of the previous researcher. Thus, he/she develops a new methodology to solve an existing problem.

As mentioned elsewhere, researchers should unshackle their creative thought processes and develop the questioning attitude and lateral thinking. This will lead to cross breeding of ideas. Such cross breeding will help the researcher to make empirical study of such ideas generated. If the outcome of the study is repeated in different contexts and the same results are obtained, they may attempt to generalize the outcome and propose new concepts or theories. For this, the research scholar should really be motivated by the spirit of inquiry and driven by higher intellectual pursuit.

There may be apprehensions in the mind of a researcher that too much of originality may lead to questioning of some established facts, and that the evaluators may hesitate to accept. But, however this should not dampen the spirit of the researcher if his/her extraordinary intellect is able to help in to putting up a convincing argument backed up by empirical evidence. Of course, this needs the whole-hearted support of his/her mentor.

Perhaps, the easiest way may be to go through the theses of previous researchers and find out the scope of further work identified by them in their theses. All that a research scholar has to do is to ensure that other scholars have not done or are not pursuing the same approach. Even if somebody deals with the same problem, he/she should make sure that the approach they propose to adopt is entirely different from others. May be, publication of a review paper indicating the method he/she proposes to follow will enable the research scholar to reinforce confidence in his/her approach.

2.2.3 A Suggested Approach to Identify a Research Problem

It appears that there is no universal approach to identify the research problem especially as how to start defining the research topic and how to zoom in on a specific research problem. The following are general guidelines based on the suggestions made by Ahmed Helmy. These could be followed by research students of M.S. and Ph.D. programmes.

i. First, a direction or area of interest is chosen and this is based on the background knowledge of the researcher obtained through courses undergone, readings made, conferences attended and discussions held with professors and peers, etc. There could be a list of topics of interest to the researcher, but it could be limited to not more than three.

ii. A set of "keywords" is compiled to start searching for higher quality readings for each of the previously selected topics. May be, one can refer to an on-line library or search on the web. But, one should check the publication details for quality.

iii. Out of the search hits, 15–20 papers are selected, which are most related to what the researcher had in mind and are of highest quality. One need not read all these papers. The title, abstract, names of authors, their affiliations and most importantly the conference or journal are checked.

iv. The set of keywords are refined and multiple searches are performed to identify most related quality work. To check for quality one can use references and citations. Usually the work most cited by high-quality papers is also of high quality. The list of references of a specific paper will provide a good direction to follow.

v. For the selected 15–20 papers, only the Abstract, Introduction and Conclusion are read in detail. If necessary the rest of the paper is skimmed through to get a general idea. The emphasis of each paper is thus identified. This emphasis could be on
   - which problem it addresses,
   - what solution it proposes,
   - how the solution differs from previous solutions, and
   - what the major contributions and conclusions are.

vi. After going through and understanding these 15–20 papers, four or five papers are selected, which are of highest quality and address the researcher's interest and challenges in the field most appropriately.

vii. These four or five papers are read from beginning to end and the following are identified.
   - major approaches
   - methods of analysis
     - metrics
     - evaluation tools
     - analysis and interpretation of resulting measured data
   - conclusions.

viii. Based on the study, a list of what the authors may have missed in the paper/ study, gaps or limitations that could be improved upon and any ideas on how to accomplish these improvements is prepared. One can ask these questions.

- Did all/some papers use similar approaches?
- Have they used the same evaluation criteria or method of analysis?
- If not, then what are the strengths/weaknesses of each method?

ix. Also, a list of ideas that one wants to explore further is prepared. Or the background material that one wants to brush upon is prepared. This will create another list of readings in later stages.

x. A two-page tentative research proposal defining as clearly as possible the following items is written.

- Motivation
- Research challenges
- Overview of existing work
- Limitations of existing work
- Potential directions and ideas for improvement
- Expected results and impact on the field.

xi. The two-page report can be reviewed by knowledgeable friends and by the research supervisor. Based on their comments, the proposal may be rewritten. If some major items are missed, then one goes back to the 15–20 list and another 3–5 good papers are selected and, read in detail, and parts of the proposal are rewritten.

2.2.4 Other Sources to Look for Research Problems

Many industries and government organizations indicate the research areas of interest to them. By going through websites of government organizations, the thrust areas or specific topics of interest can be identified. The Government of India invests significant resources in research and development to meet the needs of various sectors such as defence, space science, etc. In every five year plan substantial provisions for research and development are made. Besides, different ministries indicate the thrust areas of research in their website and invite proposals. For example, it is reported in newspapers that during the 11th five year plan period, the Government of India ₹ as planned to spend 5300 crores for research in the emerging energy technologies, the areas including,

- Research on clean coal technology
- Carbon capture and storage
- Integrated gasification and combined cycle project
- Biofuel and hydrogen energy

- Electric and hybrid vehicles
- Production of silicon crystals
- LED manufacture

Likewise other ministries and state-controlled organizations such as DRDO also indicate the thrust areas in which more research needs to be carried out.

Similarly, annual reports of companies do provide information about their R&D activities, from which some useful hints can be picked up and explored. One can find many excellent research topics by considering specific products or processes already in existence and looking into ways of better understanding their behaviour and improving their performance. When properly defined, this is not "routine testing" inappropriate for academic research study but rather the essence of most experimental study in engineering. Many industrial undertakings may be willing to contribute by way of sample products and company engineers willing to share their expertise.

In many standard books which deal with the broad area of research interest, the authors would have indicated the current status and what further exploration needs to be made. It is better to go through the latest edition of such books to identify possible specific areas, which can be explored.

It is a normal practice for researchers to indicate in their Ph.D. theses, the scope for further research and the limitations of their study due to certain constraints. By going through these theses, one can identify many research problems for investigation.

Discussion with experts who have been working in the proposed area for several years and attending of seminars, conferences and workshops related to the areas of interest to the researcher will provide ample opportunities to choose a topic of research.

2.3 PROBLEM DEFINITION

The selection of a topic for research obviously includes a certain level of definition of the problem selected. It is imperative to develop a thoughtful and detailed problem description early. Quite often one hears "A problem well defined is a problem half solved". The problem to be investigated must be defined unambiguously, for that will help to discriminate relevant data from the irrelevant ones and also identify the type of approach we have to adopt. It helps to answer questions like,

- What are the data to be collected?
- What are the relevant characteristics of the data that need to be studied?
- What are the relations that are to be explored?
- What are the techniques, experiments, etc. that are to be used?

In an academic environment, the choice of the problem topics can be wider since no economic pay-off is needed. But, however, the major constraints are the available time frame, facilities and preparation for studies. The faculty, who is a guide to the researcher, is likely to exert influence on the choice of topic.

Defining a problem involves the task of laying down boundaries within which a researcher shall study the problem with a predetermined objective in view. Though this is a little difficult task, it has to be tackled intelligently to avoid the perplexity encountered in the research operation. The usual approach of researchers posing questions to themselves and setting up techniques and procedures for throwing light on the questions concerned for defining the research problem may not produce definite results. Defining a research problem properly and clearly is a crucial part of research study and must in no case be accomplished hurriedly.

The research problem should be defined in a systematic manner giving weightage to all related points. A suggested technique for the purpose involves the following steps generally one after the other.

i.   State the problem in a general way

ii.   Understand the nature of the problem

iii.   Survey the available literature

iv.   Develop ideas through discussions

v.   Rephrase the research problem into a working proposition

2.3.1 General Statement of the Problem

Keeping in view either some practical concern or some scientific or intellectual interest or taking the clue from other published sources, the problem is stated in a general way broadly. The researchers should thoroughly immerse themselves first in the subject matter, which is concerned with the problem of study. Sometimes, the guide gives a broad hint, and it is then up to the researcher to narrow it down and phrase it in operational terms. There may be some ambiguities in the problem stated, which must be resolved by cool thinking. Also, the feasibility of a particular solution has to be considered and the same should be kept in view while stating the problem. If necessary a pilot survey or cursory simulation could be attempted.

2.3.2 Understanding the Problem

The next step in defining the problem is to understand its origin and nature clearly. The best way of understanding the problem thoroughly is by going through benchmark or base papers, connected books and other literature. The researcher can discuss with other researchers who worked on similar problems or with persons who have a good knowledge of the problem concerned. The researcher should also keep in view the environment and facilities within which the problem is to be studied.

2.3.3 Surveying the Available Literature

Study of available literature concerning the problem broadly considered will be very useful for identifying the type of difficulties that may be encountered in the proposed approach as also the possible analytical shortcomings. At times such studies may also suggest useful and even new lines of approach to the problem being considered.

2.3.4 Develop Ideas Through Discussions

Discussion concerning a problem often produces useful information. As already stated, researchers should discuss their research problem with colleagues or others who have enough experience in the same area or are working on similar problems. Such discussions should not be confined to formulation of the specific problem at hand, but should also be concerned with the general approach to the given problem, techniques and tools that might be used, possible solutions, etc.

2.3.5 Rephrasing the Problem

The researcher must rephrase the research problem into a working proposition. Once, one goes through the steps stated above, rephrasing the problem into analytical or operational terms is not a difficult task. By rephrasing, the researcher puts the problem in specific terms as far as possible so that it may become operationally viable and may lead to suggested solutions of a research problem, which may or may not lead to the real solutions. These propositions should be amenable to testing within a reasonable time and should be consistent with most of the known facts.

Apart from the above, the following points should also be taken into account while defining a research problem.

- Technical terms, words which have special meanings used in the statement of the problem should be defined.
- Basic assumptions relating to the research problem should be clearly stated.
- The criteria for the selection of the problem should be stated.
- The source of data, such as survey or experiment, etc. and the time period should also be considered in defining the problem.
- Scope and limitations of the problem should be mentioned explicitly.

Two examples of problem formed for doctoral study are given as case studies in the section following. The first case deals with insulation characteristics of silicone and EPDM polymer blending for high-voltage applications. This thesis was submitted to Anna University, Chennai. The second case deals with detection of predetermined tool wear and tool wear characterization, a thesis submitted to IIT, Madras.

Case Study 1

Insulation Characteristics of Silicone and EPDM Polymeric blends for High Voltage Applications

Polymeric housing materials in insulators are being increasingly accepted for use in outdoor applications. Nearly 40% of the recently installed insulators in developed countries (including US) are polymeric insulators. This is due to their advantages over the traditional ceramic and glass insulators such as light weight, higher mechanical strength-to-weight ratio, resistance to vandalism, better performance in the presence of heavy pollution in wet condition and better voltage-withstanding capability. But, they possess different physico-chemical characteristics. Hence, a better understanding is required so that their wider use in outdoor applications can be made reliable. Moreover, polymeric insulators are relatively new, their expected lifetime and long-term reliability are not known and therefore are of concern to users.

Preliminary literature surveys made indicate that not many research studies of this type have been attempted in the recent past. Moreover, the proposed study is an interdisciplinary research work, involving Electrical Engineering and Polymer Technology.

The lead polymer that is currently used in high-voltage outdoor applications is silicone rubber. Silicone rubber has excellent dielectric properties coupled with high temperature stability. However, silicone as a material suffers from poor mechanical strength and also involves high cost, especially in countries like India. Ethylene Propylene Diene Monomer (EPDM) elastomer when added to silicone rubber imparts good mechanical strength and outstanding resistance to attack by oxygen, ozone and weather. Developing a cost-effective polymeric blend for insulators and gaining insights on their performance is an important objective of this research study.

Blending of these two polymers, silicone rubber and EPDM, is an attractive way to develop a new housing material with good dielectric characteristics, thermal stability and resistance towards polluted environment. However, it is important to identify the optimum blend of silicone rubber and EPDM that possesses the salient advantages of both the component materials. With this objective, an extensive research is carried out to understand various electrical and mechanical characteristics of the silicone rubber first, secondly EPDM, and then a blend of these two materials. The proposed research aims to identify the best blend of silicone rubber and EPDM and test these blends for their electrical and mechanical characteristics as per the IEC (International Electro Technical Commission) and ASTM (American Standards for Testing Methods) standards.

The role of fillers has gained considerable importance in the recent past as fillers impart thermal stability to the materials and thereby improve the tracking resistance. The polymer product can have filler loading levels in the range of 20% to 80% their weight. The advantages of silica filler are better reinforcement, good thermal stability, improved service temperature, improved dielectric properties, etc. The advantages of ATH filler are non-smoking, low toxicity, being halogen-free and flame retardant. Use of alumina trihydrate (ATH) helps in avoiding surface carbonization. Instead it leads to gasification and thus provides resistance against tracking. In this work, the influences of adding fillers like silica and alumina trihydrate on the electrical and mechanical characteristics of the best blend of silicone elastomer and EPDM will be investigated.

There are few studies to establish the efficiency of surface-treated filler (e.g., silane-treated silica flour). It is expected that a better dispersion of the filler in the bulk polymeric blend could enhance the efficacy of the filler. Also, the silane addition to the polymeric blend could increase its degree of hydrophobicity. So, the study also envisages exploring the effect of silane treatment of the fillers added to the best blend of silicone rubber and EPDM.

The failure of outdoor insulation structures is mainly due to tracking. Assessing the tracking resistance is time-consuming and is an important performance index of the polymeric insulator. Hence, a relatively simple method in the form of an empirical correlation is proposed in this work, to predict the tracking resistance of polymeric insulators, in terms of other electrical characteristics which are relatively easier to measure.

The mechanical properties like tensile strength of an insulator also play an important role in determining its efficacy. Empirical correlations have been proposed to predict tracking resistance and tensile strength of a polymeric insulator based on dimensional analysis (DA) technique and multivariate regression analysis. These correlations will relate the key electrical characteristics of an insulator to its tracking resistance and tensile strength. The correlations and the associated qualitative interpretations attempted in this work will help to gain deeper insights into the characteristics of polymeric blend with and without fillers.

Case Study 2

Detection of Predetermined Tool Wear and Tool Wear Characterization

In automated manufacturing, continuous monitoring of tool wear is essential as worn tool leads to part rejection. The existing techniques of tool wear monitoring can be classified as direct or indirect. In direct tool wear monitoring, the tool wear parameter is directly measured. In indirect methods, one or more parameters like tool vibration, cutting force, tool tip temperature or acoustic emission, which are influenced by tool wear, are monitored.

Though direct tool wear methods are reliable, they are usually offline and intrusive. That is, the machining has to be stopped for carrying out the measurement. The indirect methods can be used on-line without disturbing the machining process but since the measured parameters are influenced by many other factors like cutting conditions and tool and work material properties, the reliability of learning about the tool wear status is low. The present trend is to improve the reliability of indirect tool wear monitoring by multisensor approach and improved signal processing.

Poor reliability and frequent false alarms force many manufacturers to shun away from monitoring systems. For useful implementation, the tool wear monitoring system must have the accuracy of direct monitoring and the possibility of on-line implementation. Hence the objective of the work is to develop a reliable and direct system for the detection of the onset of a predetermined amount of tool flank wear in turning.

In the present work, the gross shape of the cutting tool is proposed to be modified to have a sensing step in its flank face. The sensing step will be designed in such a way that, up on the flank wear of the cutting tool reaching a predetermined value, the sensing step would contact the workpiece, leading to its fracture. This fracture could be detected using acoustic emission monitoring. Since the cutting edge is not disturbed, the cutting can continue to finish the pass and the tool can be subsequently changed.

To improve the reliability of detection, the root mean square (r.m.s.) value of the acoustic emission signals will be computed with a time constant of 0.1 ms. The r.m.s values at the time of sensing edge fracture will be compared with the r.m.s. values observed during normal turning.

2.4 LITERATURE SURVEY

2.4.1 Importance of Literature Survey

Once we have agreed upon a clear problem definition, the next logical step will be survey of background literature. This is essential whether the proposed problem deals only with theoretical or with experimental aspects or both. The survey should be extensive which means that all available literature concerning the problem at hand must necessarily be surveyed and examined before the definition of the problem made. The researcher must be well-conversant with concerned theories in the field, reports and records and all other relevant literature. The researcher must spend sufficient time in reviewing of literature on research studies already undertaken on related problems.

The literature survey is mainly to find out what material and other data are available for operational purposes. If somebody has done experimental study or simulation, he/she might have mentioned the problem and difficulties faced by them in their study, which may be useful for the current study. Knowing what data are available often serves to narrow down the problem itself as well as the technique that might be used.

Such literature survey would also help a researcher to know if there are any gaps in the theories, or whether the existing theories applicable to the problem under study are inconsistent with each other, or whether the findings of the different studies do not follow a pattern consistent with the theoretical expectations and so on. All these will enable researchers to take new strides in the field of furtherance of knowledge, i.e., they can move up starting from the existing premise. Studies on related problems are useful for indicating the difficulties that they may encounter in the present study and also the possible analytical shortcomings. At times such studies may also suggest useful and even new lines of approach to the present problem.

2.4.2 Sources of Information

The sources of information can be classified as indicated below.

1. Public sources
    i.   Central Government Departments (Defence, Energy, Science and Technology, etc.)
    ii.   State and local Government (Highways, Pollution Control Board, etc.)
    iii. Libraries
    iv.   Universities
    v.   Internet
2. Private sources
    i.   Nonprofit organizations and services (professional societies, trade associations, membership organizations)
    ii.   Profit-oriented organizations (manufacturers, vendors' catalogues, samples, test data, etc., consultants)
    iii. Individuals (direct conversation or correspondence, personal friends, faculty)

Depending upon the nature of the problem undertaken for research study, the researcher has to seek the needed information from different sources. However, the major sources of information nowadays are the library and internet. When we look for information in the library, there is a hierarchy of information sources as given below.

- Technical dictionaries
- Encyclopedias
- Handbooks
- Bibliographies
- Indexing and abstract services
- Technical and professional journals
- Translations
- Technical reports
- Books
- Patents
- Catalogues and manufacturer's brochures

When we deal with a project in a new technical area, there may be a need to have a broad overview of the subject. Technical dictionaries and handbooks provide useful technical data. Many handbooks also provide ample technical description of theory and its application, so they are good refreshers of material once studied in greater detail.

Some libraries and research institutes provide indexing and abstracting services. These provide current information on periodical literature and they also provide a way to retrieve published literature. An "indexing service" cites the article by title, author, and bibliographic data. An "abstracting service" also provides a summary of the content of the article. Some of the indexes and abstracts include,

Ceramic Abstracts,

Chemical Abstracts,

Corrosion Abstracts,

Fuels and Energy Abstracts,

Highway Research Abstracts,

Applied Science and Technology Index,

Engineering Index

Another useful source of much detailed information is "Dissertation abstract", which gives abstracts of most doctoral dissertations compiled in the United States and Canada. A copy of the dissertation can be ordered at a nominal cost. In India, the UGC has started including abstracts of research theses in its website. Now, the UGC has made it mandatory for all universities to send them a soft copy of the thesis.

The various sources from which information is gathered can also be grouped into (i) primary sources (ii) secondary sources and (iii) tertiary sources. Primary sources provide first-hand or original information. For example, data gathered out of an experimental study, papers of original authors wherein no distortion or no screening of information occurred are all primary sources. The secondary sources of information contain information culled out from primary sources and these include abstracts, review articles, etc. The information provided by them may not be complete as the authors might have abridged them or screened them to suit their requirement. Textbooks, handbooks, etc. are examples of tertiary sources of information. The information contained in them might have been taken from primary and secondary sources. Veracity of such information depends on the credentials of the author, publisher and source of authenticity.

2.4.3 Conducting Literature Surveys

Conducting a search of the published literature is like putting together a complex puzzle. More than three million publications are brought out every year. With the development in ICT, extensive computer-aided databases are available to the researcher from which he/she can retrieve the required information. Nowadays, on-line data search is primarily employed through internet. In the on-line mode, a direct dialog between the searcher and the host system takes place through network using search engines. One has to select a starting place, but some starts are better than others. A good strategy is to start with the most recent subject indexes and abstracts and try to find a current review article or general-purpose technical paper. The references cited in it will be helpful in searching back along the "ancestor references", to find the search that led to the current state of knowledge. However, this search path will miss any references that were overlooked or ignored by the original researchers.

The next step should therefore involve citation search to find the descendent references using "Citation Index". Once the researcher finds a reference of interest, he can use citation index to find all other references published in a given year that cited the key reference. These two search strategies will uncover as many references as possible about the topic.

Citation index (CI) is an index based on the principle of existence of some meaningful relationship between a paper and some other paper that it cites or is cited by the other. Thus, it establishes a relationship between the works of two authors or two groups of authors who published papers. In fact, the entire database of science citation index is amenable to analysis by extensible manipulation. It is also possible to identify the frequency authorwise, with which their name or their papers are cited in the literature over any selected time frame. All these counts are quantitative and objective.

By taking note of the author's or his/her research paper's frequency of citation, the author's professional standing can be established through correlation. The citation index provides a very useful and an objective criterion for judging the quality of a paper published in a journal. On the same analogy, one can evaluate a journal on the assumption that if a particular journal article is frequently cited, that journal has a high impact factor.

The next step is to identify the key documents. One way to do this is to identify the references which have the greatest number of citations, or those that other experts in the field cite as particularly important. One should keep in mind that it takes 6 to 12 months for a reference to be included in an index or abstract service, so current research will not be picked up using this strategy. Current awareness can be achieved by searching current context on a regular basis using keywords, subject headings, journal titles and authors already identified from literature search.

2.4.4 Assessment of Quality of Journals and Articles

The quality of journals in a discipline/area can be assessed by using "Impact factor". The impact factor of a journal is calculated by dividing the references cited in one year by the number of citable articles published in the same journal over the previous two years. This ratio is published annually in Thompson Scientific Journal Citation Reports (JCR) along with a number of other quantitative tools for comparing and evaluating scholarly journals. There are many databases such as Web of Science, SCOPUS, Google Scholar, Publish or Perish. While Web of Science or SCOPUS are subscriber databases, Publish or Perish and Google Scholar are free-wares.

The scientific Journal Citation Report (JCR) impact factors are all based on data from Journals indexed in Web of Science, but the SCOPUS uses a measure called "h-index", which was developed by Hirsch in 2005 to evaluate the impact of journals. Subsequent comparative studies made by researchers indicate that the JCR impact factors are by and large close to SCOPUS impact factor. The h-index is an index that attempts to measure both the scientific productivity and the apparent impact of scientists. This is based on the set of the scientist's most cited papers and the number of citations that they have received in other people's publication.

The calculation of h-index can be explained as follows: A scientist has index 'h' if 'h' of his/her 'N' papers have at least 'h' citation each, and the other (N–h) papers have almost 'h' citation each. This h-index serves an alternative to more traditional journal impact factor metrics in the evaluation of the impact of the work of a particular researcher. One can calculate the h-index using freely available databases—Google Scholar and Publish or Perish. The subscriber databases such as SCOPUS and Web of Science provide automatic calculation. The h-index can also be used to evaluate the productivity and impact of a group of scientists, such as those in a Department or University or even a country.

Science Citation Index (SCI) is a citation index originally produced by the Institute for Scientific Information (ISI) in 1960, and now owned by Thompson Reuters. This has been expanded into a larger version "Science Citation Index Expanded" and it covers more than 6500 notable and significant journals across more than 150 disciplines. These journals are the world's leading journals of science and technology and the selection process is very rigorous. It is available through Web of Science database, a part of Web of Knowledge collection databases.

While the quality of a journal is judged by the impact factor of the journal, the quality of an article is established by the number of citations, h-index, etc. In fact, the University Grants Commission and Universities attach significant importance to all these in judging credibility of scientist, Institution and articles published.

2.4.5 Information Through Internet

The fastest growing communication medium is the Internet. The internet is a computer network interconnecting numerous computers or local computer networks linked by common technical protocols. The most recent and most rapidly growing component of the internet is the World Wide Web (www), an enormous, far-flung collection of colourful on screen documents that are linked to each other by highlighted words called hypertext. The popularity of www comes from the fact that it makes distribution and accessing digital information simple and inexpensive. Since a large collection of documents are involved, using the web requires a "search engine". Nowadays a number of user-friendly highly graphical search tools, often called "web browsers" are extensively used. The most commonly used include Netscape Communicator and Microsoft Internet Explorer.

Locations on the internet are identified by Universal Resource Locators (URL). For example, a URL that gives a brief history of the Internet is http:/www.isoc.org internet-history. The prefix http://www indicates we are typing to access a HTTP server on the World Wide Web at a computer with the domain name "isoc.org". The document is stored in that computer in a file called "internet–history".

To search the World Wide Web, we need a search engine. A search engine usually refers to a web search engine, which searches for information on World Wide Web (www) through Internet. There are other kinds of search engines such as enterprise search engines, which search on intranets, personal search engines, mobile search engines, etc. Each search engine works differently in combing through the web. Some scan the information in the title or header of the document, while others look at the bold headings on the page. In addition, the way information is sorted, indexed and categorized differs between search engines. Therefore all search engines will not produce the same result for a specific enquiry. The most commonly used general-purpose search engines include,

- Alta Vista   http://www.altavista.digital.com
- Google Scholar   http://scholar.google.com
- Microsoft Academic   http://academic.research.microsoft.com

Search engines like Scirus, Sci Net, Scholar provide for most comprehensive scientific and technology information. These search engines perform the following steps, viz.,

    i.   Crawling

   ii.   Indexing and

  iii.   Searching.

It is important to understand how search engines obtain and maintain their databases and also the search methods. Another important aspect is the degree to which a search engine covers all the materials out of the web. A recent study estimated that even the best search engine covered about one-third of the web pages only. Therefore, the best strategy for maximizing information retrieval from internet is to use two or three search engines.

The researcher should also have knowledge of many URLs devoted to engineering topics and the need of the user. Some examples of URLs are,

NASA Technical Information Service	http://tech reports.lors.nasa.gov/egi-bin/ NTRs. This covers all NASA centres and provides abstracts only. Reports must be ordered.
National Technical Information Service	http://ntis.gov. Large number of reports are available on file.

The purpose of a literature search are several. The main purpose is of course, to find whether the study you are about to undertake has already been done. We do not want to waste time and energy repeating someone else's work, unless we question their results for some reasons. Because of the vastness of open literature and the inaccessibility of many proprietary files, we can never be sure that we have uncovered all works that have been done. Therefore, a sincere effort in doing a thorough literature survey is essential.

2.5 LITERATURE REVIEW

2.5.1 Need for Review

The purpose of literature review is primarily to demonstrate the level of understanding of the researcher related to his/her project. It should cover specifically studies of literature, which are of benchmark type pertaining to the area of research. The major reasons for providing a literature review include the following.

- To defend the researcher's choice of problem undertaken for research, whether it is theoretical or experimental, and the proposed methodology.

- To what extent the research topic chosen is important and current.

- To provide background information required to ensure better comprehension of the research work and the familiarity of the researcher with them.

- To provide a proper link between the research work and the state-of-the-art prevailing in the aspect considered.

Literature review is an essential exercise by the research scholar to review the critical points of current knowledge and methodologies of approaches adopted on a particular topic of interest to the researcher. The literature reviewed comprise scholarly articles, books, dissertation, conference proceedings, etc. relevant to the topic chosen. As many of these are secondary sources they may not report any new original experimental work. The researcher should bear in mind that the literature review is neither a descriptive list of material gathered nor a summary of them. The literature collected should be relevant and should have bearing on the research objective, the research problem undertaken for study and the approach or methods that he/she has planned to adopt. The ultimate goal of such an exercise is to make the research scholar up-to-date with the current literature on the topic because this forms the basis for further research work that may be needed in the area.

2.5.2 Some Guidelines for Review

There are no universal guidelines available as to how one should proceed with a literature review. However, many suggestions and hints are provided by several authors. A well-structured literature review is characterized by a logical flow of ideas, current and relevant references, proper use of terminology and an unbiased and comprehensive view of previous research on the topic. The following points will be useful to scholars while pursuing literature review.

- Ensure whether the literature gives an overview of the subject under consideration which satisfies the objectives of the reviewer.
- Categorize the literature into groups', viz., work which support a particular position, work which takes a contradictory position and work dealing with alternative approaches.
- The grouping should take into account the context of the literature's contribution to the understanding of the subject.
- Establish the credentials of the author(s) of the literature and see whether the author's arguments are supported by evidence such as specific case studies, recent findings, etc.
- Observe whether the perspective of the author is impartial or whether contrary data is considered or certain pertinent information is ignored in order to prove his/her point of view (biased approach).
- Check whether each of the literature reviewed provides similar viewpoints and how it differs from others.
- Examine if the arguments of the author and findings are convincing and that the literature contributes significantly to the researcher's understanding of the issue under consideration.
- Understand the interrelationship of each literature reviewed with other literature considered.
- Try to find new ways to interpret and also identify any gaps in the previous research related to the area of interest.

- If there are any contradictions among the reported literature on a particular issue, introspect as to how such conflicts can be resolved.

- Ensure whether the literature helps to understand the state-of-the-art of the issue being considered and how it will help to move forward in the research study.

- Check if the literature reviewed helps to establish the originality of the research study undertaken in the context of existing literature.

2.5.3 Record of Research Review

It will be very useful while preparing the thesis if a proper record of review of literature gathered is made by the researcher. The review should encompass genesis of the problem in the researcher's mind, the extent of knowledge already available and a fairly extensive list of significant papers or other works in the area along with summary of the relevant findings. An extensive literature survey and rigorous review lend credibility to the problem formulation by the researcher for his/her research study. It is imperative that the researcher sums up as to how the literature relates to his/her research problem. This can be done easily if the researcher keeps a proper recor of these. A suggested form of literature survey/review is given below.

Author(s)	Paper title	Publication details	Points	Conclusion and future work
Hung-Yu- Chin	Secure Access Control schemes for RFID system with anonymity	Mobile Data Management, 2006. (MDM2006). 7th International Conference. 10–12 May 2006. pp 90–96	Forward vs Backward channel Hash-based access control Randomized access control	New Hash scheme proposed

The above format is only indicative, however, the research scholars using their ingenuity can design a proper format for storing review data. In this context, it is suggested that they can make use of Excel's spreadsheet to maintain a database of literature review, otherwise they can use "Access" the regular database program included in the "MS-office". In fact, Excel spreadsheet provides several tools for working with data lists. The researcher can use a data entry form to enter data, edit, sort and display any subset of information and more. Apart from tracking data stored, they can add, delete and modify information stored.

Apart from database features, the Excel spreadsheet provides a lot of facilities for creating "worksheets", where one can enter formulae or numerical data, manipulate data to make several types of calculations. Various functions like algebraic, statistical, financial, etc. are also available in the Excel worksheet. With graphics facilities available, the researcher can prepare various charts such as bar and Pie charts and graphs. The spreadsheet software will be a very versatile one for the researcher.

It will be very useful if the researchers, based on the literature survey and all other materials they have gathered and critically examined, prepare a review report bringing out the state-of-the-art situation about the topic they have chosen. They should also identify the gaps and how their topic of research will be a useful addition to the literature. Based on this they can prepare a review paper and publish it. Publication of a review paper will help to infuse confidence in their mind about the problem they propose to explore. They can use the review details to present papers in conferences. Moreover this will provide adequte material for writing the chapter on "Literature review" in the thesis.

3
Research Design

3.1 RESEARCH DESIGN: WHAT IT IS

After formulating and defining a research problem and deciding that it is worth investigation, supported by evidences from literature survey and review, the researcher has to prepare an action plan as how to proceed. This action plan contemplates dealing of issues such as how to gather data, from what source(s) and conditions, tools and techniques used, and whether through laboratory experiment or simulation or by means of survey appropriate to the research study. Besides, the plan should cover methods to be used for analysis of data gathered. The intended purpose of the plan is to ensure the creditability of the proposed study. In other words, the researcher has to evolve a "research design", as it is nothing but an arrangement of conditions for collection and analysis of data relevant to the objective and scope of the study coupled with economy. In fact, it is a blueprint for the proposed study.

The overall research design can be split into the following groups of activities:

i. Evolving a suitable method for selecting items to be observed for the study—sampling design.

ii. Identifying the conditions under which the observations are to be made—observational design.

iii. Methods dealing with as to how many items are to be observed and how to gather information/data and how to analyse them—statistical design.

iv. Identifying the techniques for carrying out the procedures stated in (i) to (iii)—operational design.

3.2 WHY WE NEED RESEARCH DESIGN

For any project to be successful, the first and foremost requirement is a proper action plan for the project. As research work is normally a medium-term project with an objective, scope and time frame, it is imperative to have a proper research plan or research design. Not only will this ensure credibility and smooth sailing of the research operations, but also will make it efficient, thereby yielding the required data/information with minimum expenditure, effort and time. The preparation of the research design should be carried out with great care as any error in it may upset the entire project. This improves the reliability of the results arrived at as well as constitutes a firm foundation of the entire edifice of the research work.

A good research design should be flexible, appropriate, efficient and economical and should be related to the purpose or objective of the problem undertaken for investigation. As the nature of the problem differs from study to study, it is not possible to have a common design for all research studies. The design will vary from study to study, since a design suitable to a particular study may be inadequate or inappropriate in the context of another study. In fact, the research designs are problem-specific.

In an exploratory research study where the major emphasis is on discovery of ideas and insights, the research design must be flexible enough to permit consideration of many different aspects of a phenomenon. But in a study dealing with accurate description of a situation or an association between variables, accuracy becomes a major consideration. Studies involving the testing of hypothesis of causal relationship between variables require a design which will permit inferences about causality, ensuring minimum bias and maximum reliability.

We may not be able to put a particular study in a particular group in practice, as a given study may have in it elements of two or more of the functions of different studies. Only on the basis of primary function can a study be categorized as exploratory or descriptive or hypothesis testing and accordingly the choice of a research design may be made in the case of a particular study.

3.3 TERMINOLOGY AND BASIC CONCEPTS

We use various terms and concepts in research design. Some of the important terms with short explanation for each of them is given to ensure better understanding of the subject.

System An assemblage of objects, ideas or activities interrelated in some way, which permits identification as a coherent whole; a logical or functional unit. A system is conceived for a specific purpose.

Concept A visionary expression of a proposed or planned action that leads to achievement of a disciplinary objective.

Principle A guiding thought or fundamental truth based on empirical deduction of observed behaviour or practice that proves to be true under most conditions over time (e.g., Archimedes principle).

Law   Laws are formulated or propounded in particular instance based on regularity in natural or other occurrences (e.g., law of gravity, law of conservation of energy).

Theory   Facts or a system of ideas based on general principles which are capable of explaining a phenomenon or occurrence of a thing but not depend on the particular phenomenon or thing explained (e.g., Theory of evolution, Theory of elasticity).

Bias   Bias implies prejudice and there are various kinds of bias depending on the context. In statistics, the word "bias" refers to systematic prejudice that is present in the data collection process or while taking readings using instruments with defects. Sampling bias occurs when some members of the population are more likely to be included than others. Systematic or systemic bias refers to external influences that may affect the accuracy of statistical measurements.

Process   A sequence of serial and/or concurrent operations or tasks that transform and/ or add value to a set of inputs to produce a product. Processes are subject to external control and constraints imposed by regulators and/or decision authority.

Practice   A systematic approach that employs methods and techniques that have been demonstrated to provide results that are generally practicable and repeatable under various operating conditions.

Simulation   The process of designing a model of a real or a hypothetical system and conducting experiments with this model for the purpose of either understanding the behaviour of the system or of evaluating various strategies for the operation of the system.

Variable   An entity which can take on different quantitative values. For example weight, stress and strains are variables. Qualitative phenomena (or attributes) are also quantified on the basis of the presence or absence of the concerned attribute.

Continuous and discrete variables   Certain variables take continuous values and they are called continuous variables and variables which take only fixed, integer values are called discrete variables.

Dependent and independent variables   If one variable depends upon or is a consequence of another variable, it is termed as dependent variable; and the variable that takes values on its own and is not dependent on another variable is called independent variable.

Input and output variables   In an experiment some of the independent variables are considered as input variables by the experimenter and the variable, whose outcome is of interest is called output variable or response variable. Input variables are also called "factors".

Extraneous variable   Independent variables that are not related to the purpose of study, but may affect the dependent variable are termed as extraneous variables. Any effect noticed on the dependent variable as a consequence of extraneous variable is technically called as experimental error.

Control   One important characteristic of a good research design is to minimize the influence or effect of extraneous variable(s). The technical term "control" is used when we design the study minimizing the effect of extraneous variables.

Confounded relationship   When the dependent variable is not free from the influence of extraneous variables, the relationship between dependent and independent variable is said to be confounded by an extraneous variable.

Research hypothesis   A prediction statement that relates an independent variable to the dependent variable.

Experimental and non-experimental hypothesis testing   When the purpose of research is to test a research hypothesis, it is termed as hypothesis-testing research. Research in which the independent variable is manipulated is termed as experimental hypothesis-testing research. On the other hand, a research in which an independent variable is not manipulated is called non-experimental hypothesis-testing research.

Experimental and control groups   In an experimental hypothesis-testing research, when a group or sample is exposed to usual conditions, it is termed as control group, but when the group is exposed to some novel or special conditions, it is termed as experimental group. It is possible that certain studies involve only experimental groups or both experimental and control groups.

Treatments   The different conditions to which experimental and control groups are usually exposed are referred to as treatments.

Experiment   The process of examining the truth of a statistical hypothesis, relating to some research problem is known as experiment. Experiments can be of two types, viz., absolute experiment and comparative experiment.

Experimental units   The predetermined plots or the blocks, where different treatments are used are known as experimental units. Such experimental units must be selected (defined) very carefully.

3.4 DIFFERENT RESEARCH DESIGNS

Research designs can be categorized as:

i.   Designs involving exploratory research studies,

ii.   Descriptive and diagnostic type research design studies, and

iii. Research studies involving hypothesis testing.

3.4.1 Design Involving Exploratory Research

This type of study is also termed as formulative research study. The major emphasis of such studies is the discovery of concepts (ideas) and better insights. Since, this is of exploratory nature, the study design should be flexible enough to provide opportunity for considering different aspects of the problem under study. In this, the research problem is initially defined broadly, and may get transformed into one with more precise meaning. This may require changes in the research procedure for gathering data, hence in-built flexibility in research design is needed.

In studies relating to engineering, the experimental work may be tentative and somewhat of unfocused nature initially or in the early stages of familiarization with physical phenomena. Hence, one needs to keep an open mind, in the hope of observing some unusual or potentially useful behaviours, which might become the subject matter later for more focused investigation. In routine work we usually look for reasons to throw out data points that do not repeat themselves (outliers); but in exploratory work the "wild" point may be the basis of an important new product or process, and hence we should not be too quick to dismiss it.

For exploratory research work, the following three methods of research designs are contemplated i) survey of concerned literature, ii) experience survey, and iii) analysis of insight-stimulating examples (i.e., typical case studies).

Survey of concerned literature   The survey of literature concerning the issue is the most simple and fruitful method of formulating the research problem or developing the necessary premise. The premise or hypothesis stated by earlier researchers may be reviewed and their usefulness evaluated as a basis for further research. Thus, the researcher should review and build upon the work already done by others, but in cases where no premise has been formulated, the task of the researcher will be to review the available material for deriving the premise from it.

Apart from the bibliographical survey of studies made in one's area of interest, the researcher should cross-breed concepts and theories developed in different research areas to the area in which the researcher is currently working. For example many of the concepts in biology have been successfully adopted in engineering designs, optimization, etc.

Experience survey   Experience survey involves survey of people or experts who have practical experience with the problem to be studied. The object of the survey is to obtain insight into the relationships between variables and new ideas relating to the research problem. This survey may enable the researcher based on discussion with experts to define the problem more concisely and help in the formulation of a research hypothesis.

Analysis of insight-stimulating examples In a way, this is akin to case study method. This method is particularly suitable in issues where there is little experience to serve as guide. In this approach an intensive study of selected instances of phenomenon in which the researcher is interested will be useful. Further, existing records relevant to a situation or phenomenon are examined and in addition unstructured interview of selected persons with sharp contrast or having striking features are held. Mostly, this method is adopted in social-science-oriented research.

Thus, in an exploratory research study which leads to insights or premises, whatever method of research design outlined above is adopted, the only thing essential is that it must continue to remain flexible, so that many different facets of the problem be considered as and when they arise and come to the notice of the researcher.

3.4.2 Descriptive and Diagnostic Research Studies

Descriptive research studies are those studies, which are concerned with describing the characteristics of a particular material, process, individual or a group. In diagnostic research studies, we are concerned with the frequency with which something occurs or its association with something else. Studies concerning whether certain variables are associated and if so how are examples of diagnostic research studies. Investigation which leads to characterization of materials in chemistry, polymer, etc. comes under descriptive study. Most of the social research comes under this group.

From the point of view of research design, descriptive as well as diagnostic studies share common requirements. The researchers in both the cases should not only define clearly what they want to measure but also indicate suitable methods for measuring it along with clear-cut definition of the "population", they want to study. The research design must have enough provision for protection against bias and to maximize reliability, with due concern for the economic completion of the research study.

In both the cases, the design must be rigid and inflexible and should focus attention on the following.

- Determination of objective of the study so as to indicate what is the study about and why it is being made.
- Evolving methods of data collection and the techniques to be used.
- Sample size determination.
- Periodicity or frequency of collection of data.
- Manner of processing the data.
- Method of reporting the findings and conclusions.

Any equipment or instrument used for data collection should be properly calibrated and tested before conduct of study to minimize bias. After collection of data, they should be examined for completeness, consistency, comprehensiveness and reliability. Since sample design is involved, it should be tackled in such a fashion that samples yield accurate data/information with a minimum amount of research effort. Usually one or more forms of probability sampling methods like random sampling may be used. This calls for appropriate statistical operations, along with appropriate tests of significance to be carried out for drawing conclusions.

The differences between research designs involving exploratory research and that involving description/diagnosis type of research are summarized in Table 3.1.

Table 3.1 Differences between research desig for exploratory research and diagnostic research

Aspects	Exploratory	Descriptive/diagnostic
Overall aspects	Flexible design to consider different aspects of the problem	Rigid design to protect against bias and to maximize reliability
Sampling	Non-probability sampling design	Probability sampling design
Statistical aspect	No pre-planned design for analysis	Pre-planned design for analysis
Observation	Unstructured instruments for data collection	Structured and well-thought-out instrument of data collection
Operational	No fixed decisions about the operational procedures	Prior decision about operational procedures

3.4.3 Hypothesis-testing Research Studies

Hypothesis-testing research studies are generally known as experimental studies and in this the researcher tests the hypothesis of causal relationship between variables. Such studies require procedures that will not only reduce bias and increase reliability, but also permit drawing inferences about causality. Usually experiments meet this requirement and when we talk of research design in such studies, we often mean the "design of experiments".

3.5　EXPERIMENTAL DESIGNS

Design of experiments (DOE) or experimental design is the design of any information-gathering exercise where variation is present, whether under the full control of the experimenter or not. However, in statistics, these terms are used for controlled experiments.

The purpose of conducting a physical experiment or simulation and analysis of output data by a researcher is primarily to secure answers to the problem defined and hypothesis formulated. Though analysis follows the experimental results sequentially, it is imperative that a proper design of experiments be made to ensure that no complexities arise afterwards.

In addition, the researcher should introduce measures in the experimental design so as to minimize the effect of bias, variability and the experimental error. Thus experimental design assumes greater significance in research in order to minimize bias and improve reliability of results.

The study of experimental design has its origin in agricultural research and Professor R.A. Fisher's name is associated with experimental designs. Prof. Fisher found that by dividing agricultural fields into different blocks and then by conducting experiments in each of these blocks, the information gathered and inferences drawn happens to be more reliable. Today the experimental designs are being used in researches relating to phenomena of several disciplines.

3.5.1 Basic Principles

Professor Fisher has enumerated the following three principles of experimental designs.

    i.   The principle of replication

    ii.   The principle of randomization

    iii. The principle of local control

The principle of replication requires that the full experiment be repeated more than once. This is resorted to when the measurements are highly variable and uncertain. The experiment (treatment) is applied to many experimental units instead of one. By doing so, the sources of variation can be identified and the effects of treatment can be better estimated. An assumption made here is that the replications are independent, but often it is not. Conceptually replication does not present any difficulty, but computationally it does. For example if an experiment requiring two-way analysis of variance is replicated, it will require three-way analysis of variance since replication itself may be a source of variation in the data. Nevertheless, replication is introduced in order to increase the precision of a study, i.e., to increase the accuracy with which the main effects and interactions can be estimated.

The principle of randomization ensures protection in an experiment against bias caused due to the effect of extraneous factors. This principle indicates that we should design the experiment in such a way that the variations caused by extraneous factors can all be combined under the general heading "chance". In an experimental programme involving a large number of tests, it is important to randomize the order in which the specimens are selected for testing. By this, any one of the many specimens involved in the experiment have an equal chance of being selected for the test. By randomizing the allocation of experimental units, bias is reduced by equalizing the so-called independent variables or factors that have not been accounted for in the experimental design.

The principle of local control is another important principle of experimental design. In this, the extraneous factor, the known source of variability, is made to vary deliberately over a wide range as necessary and this needs to be done in such a way that the variability it causes can be measured and can be eliminated from experimental error. This means that we should plan the experiment in a manner that we can perform a two-way analysis of variance, in which the total variability is divided into three components, viz., attributed to treatments, the extraneous factor and experimental error. Here, the experiment is divided into homogeneous blocks, each block subdivided into parts equal to the number of treatments and treatments are randomly assigned to these parts of a block. In general blocks are the levels at which we hold an extraneous factor fixed, so that we can measure its contribution to the total variability of the data by way of two-way ANOVA. Thus, through the principle of local control, we can eliminate the variability due to extraneous factor(s) from experimental error.

3.6 IMPORTANT EXPERIMENTAL DESIGNS

As already stated use of appropriate experimental design ensures several benefits to the researcher. The most important benefit from statistically designed experiments is that more information per experiment will be obtained than with unplanned experiments. The second benefit is that statistical design results in an organized approach to the collection and analysis of information. The third advantage is the credibility that is given to the conclusions of an experimental programme when the variability and sources of experimental error are identified. Finally, an important benefit of proper design of experiment is the ability to discover interaction between experimental variables. A major impetus to the experimental design has been given by the strong emphasis being shown in the design for quality (DFQ).

The experimental designs can be broadly divided into two categories, viz., i) Informal experimental designs and ii) Formal experimental designs. Informal experimental designs are those designs that normally use less sophisticated form of analysis mainly based on difference in magnitudes. Formal designs or statistically designed experiments offer relatively better control and use precise statistical procedures for analysis.

    i.   Informal experimental designs

        a)   Before-and-after without control design

        b)   After only with control design

        c)   Before-and-after with control design.

    ii.   Formal experimental designs

        a)   Blocking designs

- Completely randomized block (CRD)
- Randomized block design (RBD)
- Latin square design (LSD)
- Graeco-Latin square design (GLD)

    b)   Factorial designs
- Fractional factorial
- Full factorial
- Use of orthogonal arrays

    c)   Response surface designs

Informal experimental designs will be discussed in this chapter, while formal experimental designs will be dealt with in a later chapter after introducing some statistical concepts.

Before-and-after without control design In this design a single test group is selected and the dependent variable is measured before the introduction of the treatment. The treatment is then introduced and the dependent variable is measured again after the treatment has been introduced. The effect of the treatment is equal to the difference between after exposure to treatment and prior to treatment. The main difficulty of such a design is that with the passage of time considerable extraneous variations may be there in the treatment effect.

After-only with control design Two groups (test group and control group) are selected in this design and treatment is given to the test group only. The impact of the treatment is assessed by subtracting the value of the dependent variable in the control group from its value in the test group. The basic assumption is that the two groups are identical with respect to the behaviour of interest. If this assumption is not true, there is the possibility of extraneous variation entering into the treatment effect. This design is superior to the previous one.

Before-and-after with control design As per this design two groups are selected and the dependent variable is measured in both the groups for an identical period of time before the treatment. The treatment is introduced in the test group only, and the dependent variable is measured in both the groups for an identical period of time. The treatment effect is determined by finding the difference between the change in the dependent variable inthe control group and the change in the dependent variable in the test group.

Details	Time period I	Time period II
Test group	Level of dependant variable before treatment (x)	Level of dependant variable after treatment (y)
Control group	Level of dependant variable without treatment (a)	Level of dependant variable without treatment (b)

$$\text{Treatment effect} = [(y - x) - (b - a)]$$

This design is superior to both the previous cases for the simple reason that it avoids extraneous variation resulting from both the passage of time and from non-comparability of test and control groups.

3.7 DESIGN OF EXPERIMENTAL SET-UP

Many research studies may involve development of an experimental set-up to meet the specific needs of the researcher. Using the experimental set-up, the researcher finds the effect of independent variable(s)on the dependent variable or response variable. Of course, the experimental set-up may vary depending upon the objective of the research study. Arranging an experiment shares many common features with product design. Design of an experimental set-up or developing a new product is a one-of-a-kind rather than a mass produced, since it is built for a specific purpose.

For developing an experimental set-up, we can follow the approaches recommended for a new product design. The flowchart for product design process is given in Figure 3.1. The whole process is divided into three major phases, viz., conceptual, substantive and detail design.

During the conceptual phase, we think of all the different ways we could accomplish the basic task the product has to perform, evaluate each approach and select those which are feasible ones and move to the next phase. In the substantive phase, we need to embody our concept in an actual physical device which can be analysed for function and cost. Each of the approach selected must be carefully analysed, improved and adapted to our specific needs and then evaluated critically against performance and cost specifications. When we reach this stage, we would have learned about the problems so that we may return to the conceptual stage and add some new concepts to our initial list. After completing the substantive design phase, one of the alternative designs stands out as clearly superior and we may reject others. If several alternative concepts have survived to this stage, an intelligent choice among them can be made.

The analogy of product design can be adopted for preparing an experimental set-up needed by a researcher for his/her study. During conceptual phase one should avoid thinking about physical hardware and should concentrate on functional requirements. The steps that may help moving from the specification to one or more concepts include:

- to identify the essential functions from the list of considerations
- establish function structures (overall and sub-functions)
- search for solution principles to fulfil sub-functions
- combine solution principles to fulil overall function
- selection of suitable combinations
- firm up concept variants

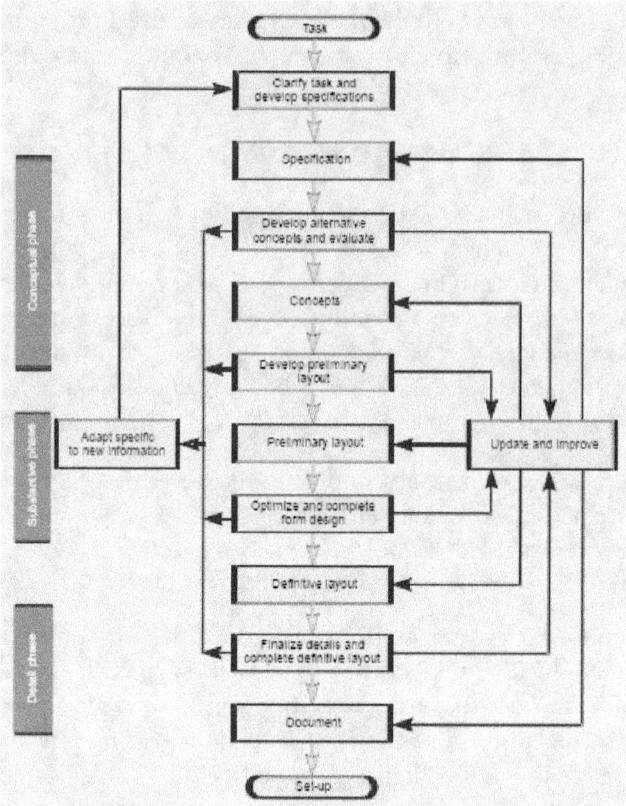

Figure 3.1 A flowchart for a product design process

The approaches that may help to search for solution principles to fulfil sub-functions are: (i) literature search (ii) known technical systems (iii) analogies (iv) previous experimental studies (v) design catalogues (vi) discussion with colleagues and experts, etc.

In the substantive or embodiment phase of experimental set-up design, the researcher should concentrate on how to convert the conceptual set-ups into physical structures or simulated structures. Here, the main emphasis is identification and assessment of equipment, instruments, sources of supply, their method of functioning, precision, cost, etc. for physical experiment; the software available, and their efficacy, limitations, etc. for simulation studies in respect of various alternative concepts developed in the first phase. Since, this being experimental set-up, the researcher has to give more weightage to the technical functioning and should ensure that nothing important to a specific set-up has been overlooked.

The final phase is a detailed planning phase. Hence, each of the alternative experimental set-up has to be thoroughly evaluated taking into account the study objective, expected outcome, feasibility, cost involved and time frame required. After evaluating each alternative the most feasible one from all considerations has to be selected for adoption in the study.

3.8 USE OF STANDARDS AND CODES

Research is a challenging task, as it deals with problems hitherto not solved. Hence, to carry out the work, it requires inputs from several sources. Apart from various sources of literature indicated earlier, the codes and standards are an important source of documents to help the researcher in the design of the experimental set-up. Standards cover instrument accuracy, test apparatus, testing methods and environment. Besides, they specify the number of specimens to be taken and the statistical tools to be used, which are useful in sampling.

Many developed and developing countries have their national standards organizations. These organizations ensure coordination and preparation of standards and codes relating to various aspects in different disciplines. While the USA has the American National Standards Institute (ANSI), the UK has British Standards (BS) and Germany has DIN standards. In India, we have the Bureau of Indian Standards (BIS) which is involved in the preparation of standards and publication for the benefit of industry and public. Internationally, the International Standards Organization (ISO) examines the standards prepared by member countries and evolves common standards.

Besides the national and international standards organizations, various professional associations are also involved in the preparation of specification, standards and codes on various aspects. For example, the American Society for Testing Materials (ASTM) prepares the annual book of standards which is an important source of information for scientists and engineers. There are several subject areas such as iron and steel, nonferrous metals, petroleum products, plastics, rubber, water and environment, instrumentation, chemicals, etc. Moreover there are other professional bodies like Society of Automotive Engineers (SAE), American Concrete Institute (ACI), Institute of Electrical and Electronic Engineers (IEEE) and Association of Computing Machinery (ACM), whose publication are worth referring. In our country, many specifications, testing procedures, etc. are formulated by Indian Road Congress (IRC) and these publications are available at a cost.

The material and information available in these standards, codes and methods provide researchers adequate information depending upon their need. When a researcher follows the standardized procedure stipulated by reputed professional bodies like ASTM, BIS, BS, DIN, etc. the credibility of the approach adopted by him/her gets enhanced.

3.9 USE OF READY-MADE APPARATUS

An important aspect is the need for the researcher to be familiar with the existing devices, which might provide all or part of the requirements for an experimental set-up needed by the researcher. This familiarity can lead to purchase of such parts or can serve as a source of idea for a part that will enable the researchers to make on their own. They can also combine and adopt features from existing devices in an optimum fashion for their specific needs. The useful sources of infomation include standardized testing procedures, vendors of testing equipment, buyer's guides, etc.

4
Mathematical Modelling

4.1 MODELS IN GENERAL

A model is defined as an abstraction or idealization of an object, system (real or hypothetical) or an idea in some form other than the entity itself. The functions of a model are several and important ones include the following.

- as an aid to thought process
- as an aid to communication
- as an aid for training and instructions
- as a tool for prediction
- as an aid to experimentation

A significant use of models is for prediction of the behaviour characteristics of the entities modelled and for conducting controlled experiments in situations where direct experiments would be impractical or prohibitive. In engineering, we come across several conceptual models like free-body diagram, electric circuit diagram, crystal lattice, etc. to improve our understanding and as a first step for further action.

A model may be either descriptive or predictive. A descriptive model enables us to understand the real-world system or phenomenon (e.g., sectional model of an aircraft gas turbine). A predictive model on the other hand is used primarily in engineering design because, it helps us to both understand and predict the performance of the system.

Models can also be classified as follows:

    i.   Dynamic or static

    ii.   Deterministic or probabilistic

    iii.   Iconic

    iv.   Analog

    v.   Symbolic

Static or dynamic models   A static model is one whose properties do not change with time; a model in which time-varying effects are considered is dynamic.

Deterministic or probabilistic models   A deterministic model describes the behaviour of a system in which the outcome of an event occurs with certainty. In many real-world situations, the outcome of an event is not known with certainty and these must be treated with probabilistic models.

Iconic models   It is a model that looks like the real thing (e.g., model of an aircraft for wind tunnel test or enlarged model of a polymer molecule). These models are used to describe the static characteristics of a system. The experimental model is a functioning model, which embodies the ideas of the design concept. It is as nearly like the proposed design as possible, but it may be incomplete in appearance. This model is subjected to extensive testing and modification. A prototype model is a full-scale working model of the design, which is technically and visually complete.

Analog models   They behave like real systems. They are used to compare something that is unfamiliar with something that is very familiar. Unlike iconic model, analog model may look nothing like a real system.

Symbolic models   They are abstraction of the important quantifiable components of a physical system. A mathematical equation expressing the dependence of the system output parameter on the input parameter is a common symbolic model. A symbol is a shorthand label for a class of objects, a specific object, a state of nature, or simply a number. Symbols are useful because, they are convenient, add to simplicity of explanation, and increase the generality of the situation. A symbolic model probably is the most important class of models, because it provides the greatest generality in attacking a problem.

Symbolic models are further distinguished between models that are "theoretical models" and based on established and universally accepted laws of nature, and "empirical models", which are the best approximation of mathematical representations based on existing experimental data.

4.2 MATHEMATICAL MODELS

A mathematical model, as already stated, uses mathematical language to describe a system. Mathematical models are not only used in natural sciences and engineering disciplines but also in social sciences. Physicists, engineers, computer scientists and economists use mathematical models extensively. The process of developing a mathematical model is termed as "mathematical modelling".

Mathematical models can take many forms, including, but not limited to, dynamical system, statistical models, differential equations, or game theory models. For example, a frequently used model in engineering is the linear multiple regresson model.

$$y = a_0 + a_1 x_1 + a_2 x_2 + \ldots + a_m x_m$$

where, $x_1, x_2 \ldots x_m$ are independent variables and y is the response variable of the system, and $a_0, a_1 \ldots a_m$ are coefficients.

In case y represents the yield strength of hot-rolled ferritic–pearlitic steel based on chemical composition, volume fraction of pearlite and the grain size, then the model developed based on experimental result is,

$$y = 13.29 + 5.30\, M_n + 10.21\, S_i + 0.22\, (\% \text{ of pearlte}) + 0.47\, \sqrt{d}$$

The term d is the mean linear ferrite intercept grain size in inches. The correlation coefficient is 0.89 and 95% confidence limit for prediction of yield is $\pm$ 3800 psi from the mean value.

Another example is a model which shows the steady-state relationship between the temperature and heat flow in a material is given by Fourier's law of heat cnduction.

$$\frac{Q_h}{A} = -k_t \frac{dT}{dx}$$

The heat transfer Q_h is related to the temperature gradient in the direction of heat flow $\frac{dT}{dx}$. The negative sign indicates that there is a temperature drop in the direction of flow.

4.2.1 Background to Modelling

Often, when engineers analyse a system to be controlled or optimized, they use a mathematical model. For analysis, engineers can build a descriptive model of the system as a hypothesis of how the system could work or try to establish how an unforeseeable event could affect the system. Similarly, in the control of a system, engineers can try out different control strategies using simulation.

A mathematical model usually describes a system by a set of variables and a set of equations that establish relationships between variables. The values of variables can be practically anything—real numbers, integer numbers, Boolean values or strings. As variables of the system they represent some properties of the system, viz., measured system outputs in the form of signals, timing data, counters and event occurrences (yes/no). An actual model consists of a set of functions that describe relations between different variables.

4.2.2 Building Blocks

There are six basic groups of variables: decision variables, input variables, state variables, exogenous variables, random variables and output variables. The grouping of variables is artificial and context-dependent. They may be overlapping also. Since there can be too many variables of each type, the variables are generally represented by vectors.

Decision variables are also known as independent variables. Exogenous variables are sometimes known as parameters or constants depending on the nature of study. The variables that are not independent of each other are known as dependent variables. For example the state variables are dependent on the decision, input, random and exogenous variables. Furthermore, the output variables are dependent on the state of the system (represented by state variables).

Objectives and constraints of the system and its users can be represented as functions of the output variables or state variables. The objective function will depend on the perspective of the model's user. Although there is no limit to the number of objective functions and constraints a model can have, using or optimizing the model becomes more computationally involved with the addition of many constraints.

4.3 MODEL CLASSIFICATIONS

Mathematical models can be classified in several ways. One such way is,

1. On the basis of their information source for development:
    i.   theoretical models
    ii. empirical models.
2. On the basis of mathematical property:
    i.   linear and non-linear
    ii.   deterministic and probabilistic
    iii. static and dynamic
    iv.   lumped models and distributed models.
3. On the basis of transparency:
    i.   white-box models
    ii. black-box models.

4.3.1 Classification Based on Information Source for Development

Whatever may be the nature of mathematical property, source of information and transparency, all models can be broadly grouped as cited in the first category, viz., theoretical models and empirical models.

Theoretical models   The term theory denotes explanation of a phenomenon which meets basic requirement about the empirical observation made, the methods of classification used and the consistency in its application among the members of the class to which it pertains. Theories are in the form of physical laws and principles. Theoretical models are developed using physical laws, material properties, equilibrium conditions, etc. to link the independent variables with the dependent variable depending on the context and requirement. These models serve as analytical tools for expository study and making predictions of the behaviour of a system.

Empirical models The word empirical refers to information gained by means of observation, experience or experiment. A central concept in science and in the scientific method is that evidence must be empirical or empirically based, i.e., dependent on the evidence or consequences that are observable by the senses. Also, this term connotes the use of working hypotheses that are testable using observation or experiment. Empirical data is the data produced by an experiment or observation. Therefore, any mathematical model developed using a set of data obtained by empirical means is known as an empirical model.

4.3.2 Classification Based on Mathematical Property

Linear vs non-linear Mathematical models are composed of variables which are abstractions of quantities of interest in the systems described and operators that act on those variables. Operators include algebraic operators, functions, differential operators, etc. If all the operators in a mathematical model present linearity, the resulting mathematical model is defined as linear, otherwise it is considered as non-linear. Exampes of linear models are,

$$L = ax + by$$

where a and b are constant and x, y are variables.

$$L = a\frac{d^2x}{dt^2} + b\frac{dx}{dt} + cx$$

where a, b, c are constants.

The question of linearity and non-linearity depends on the context, and linear models may have non-linear expressions in them. For example, a differential equation is said to be linear if it can be written with linear differential operators, but still it can have non-linear expressions in it. A common approach to non-linear problems is linearization, but this can be problematic if one is trying to study aspects such as irreversibility which are strongly tied to non-linearity.

Deterministic vs probabilistic A deterministic model is one in which every set of variable state is uniquely decided by parameters in the model and by sets of previous states of these variables. Therefore deterministic models perform the same way for a given set of initial conditions. Conversely, in a probabilistic or stochastic model, randomness is present, and variable states are not described by unique values, but rather by probabilistic distributions.

Static vs dynamic A static model does not account for the element of time, while a dynamic model does. Dynamic models are typically represented with difference equations or differential equations.

Lumped vs distributed parameters If a model is homogeneous (consistent state throughout the system), its parameters are distributed. On the other hand if a model is heterogeneous (varying state within the system), then the parameters are lumped. Distributed parameters are typically represented with partial differential equations.

4.3.3 Classification on the Basis of Transparency

A priori/subjective information Mathematical modelling problems are often classified into black-box or white-box models according to how much a priori information is available of the system. A black-box model is a system about which there is no prior information available, whereas a white-box model (glass box) is a system where all necessary information is available. Practically all systems are somewhere between the black-box and white-box models.

It is always preferable to use as much a priori information as possible to make the model more accurate. Therefore, white-box models are usually considered easier, because if you have used the information correctly then the model will behave correctly.

In black-box models one tries to estimate both the functional form of relations between variables and the numerical parameters in those functions. An often used approach for black-box models is neural network which usually does not make assumptions about incoming data.

Sometimes it is useful to incorporate subjective information into a mathematical model. This can be done based on intuition, experience, or expert opinion, or based on convenience of mathematical form. Bayesian statistics provides a theoretical framework for incorporating such subjectivity into a rigorous analysis: One specifies a priori probability distribution first (which can be subjective) and then update this distribution based on empirical data.

4.4 MODELLING OF ENGINEERING SYSTEMS

4.4.1 Some Basic Concepts

The major disciplines of engineering systems are mechanics, electricity, electronics, fluid mechanics and fluid controls, thermodynamics, magnetism, optics, etc. As mentioned already, a system is an assemblage of interrelated objects. The behaviour of a system is characterized by its response to external inputs, disturbances and initial conditions as shown in Figure 4.1. A mathematical model consisting of algebraic or differential equations is usedto characterize the relationships.

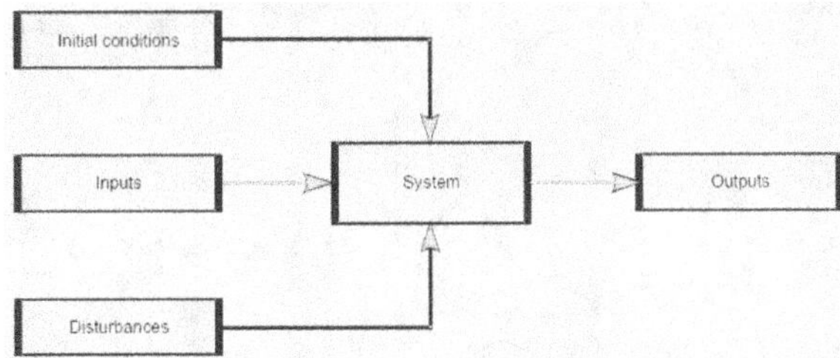

Figure 4.1 Behaviour of a system

Suppose we are using a differential equation to model a dynamic system, dependent variables of the differential equation represent the response of the system. The inputs comprise independent variables of the equation, the excitation and the forcing function to the system. By external disturbances or perturbations we mean those environmental effects that may occur randomly or unexpectedly. The initial conditions are the initial values of the dynamic variables of the system. The dynamic variables are those variables whose time derivatives appear in the governing equations.

4.4.2 Steps in Modelling

The various stages involved in the modelling of a dynamic system are:

i. Physical system of interest

ii. Modeller's perception of the system

iii. Mathematical representation

iv. Manipulation to get the response

v. Analysis of performance

Figure 4.2 illustrates the several stages involved in the modelling of dynamic systems.

As a first step, the analyst studies the actual dynamic system of interest which has all the dynamic characteristics that correspond to the exact linear or non-linear behaviour of the system. The actual system of course possesses the true response we wish to determine.

In the next step the engineer(s) perceives the system and its dynamic characteristics. The modeller may neglect some non-linearities or higher order dynamic characteristics for simplicity, even though the actual system does include all of these effects and characteristics. Hence, the engineer's perception may not truly represent the actual dynamic system.

The third step comprises formation of a mathematical model of the system represented by differential equations derived from the conservation and property laws of appropriate disciplines. If the actual system is linear, the development of a suitable mathematical model is quite straightforward. In case the system is non-linear, the mathematical model may include some approximation to simplify analysis. Therefore the equations may not represent exactly the enineer's perception of the real system.

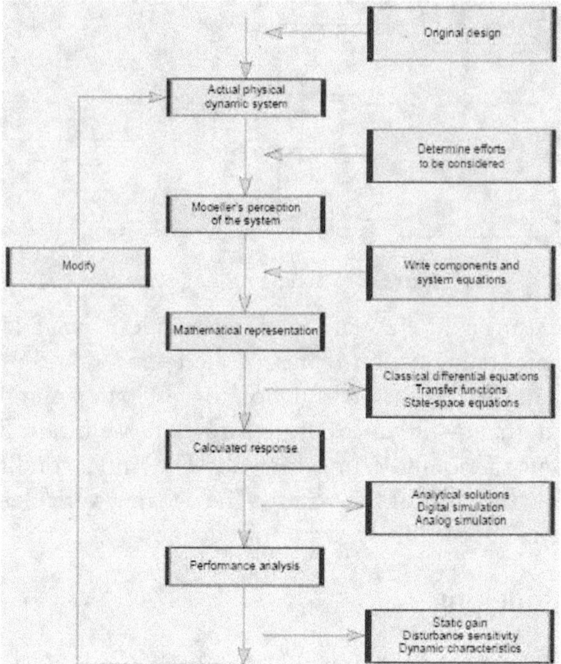

Figure 4.2 Stages in the modelling of dynamic systems

In the fourth step, the system response expressed by the analytical solutions to differential equations represents an exact solution of the equations. Some error might occur between the solution computed by the numerical method using computer and the actual solution of differential equations.

The last step involves analysis of performance of the dynamic system, as expressed by specific response measures. If the performance evaluation reveals that the system is not displaying the desired response, the system and its model should be modified or the components of the existing model adjusted to obtain the desired output.

4.5 THEORETICAL MODELS

4.5.1 Model Complexity

Mathematical models for engineering systems are developed using the conservation laws of physics and the engineering properties of each system component. These basic formulations are then combined or simplified to achieve the desired differential equation form. The preferred forms are:

i.   classical representation of a single nth order differential equation

ii.   transfer function format which gives output in terms of input

iii. state-space format of n simultaneous first-order differential equation.

Model complexity involves a trade-off between simplicity and accuracy of the model. An essential idea is that among the models with roughly equal predictive power the simplest one is the most desirable. Added complexity, though improves the fit of the model, makes the model difficult to understand and work with and can also pose computational problems including numerical instability.

Therefore, it is usual and appropriate to make some approximations to reduce the model to a manageable size. Engineers generally accept some approximations in order to get a more robust and simple model. For example Newton's classical mechanics is an approximated model of the real world. Still, Newton's model is quite sufficient for most ordinary life situations, that is, as long as particles' speed is well below the speed of light, and we study macro-particles only.

To develop the mathematical model to approximate reality, we have to make several assumptions. Even a simple context involves many assumptions. These assumptions should be rational and should not affect the accuracy of the model too much, so that the model developed serve the purpose for which it is designed.

4.5.2 Static Model

To illustrate development of a static model, let us consider the case of a beam in pure bending as shown in Figure 4.3. Pure bending implies the beam is acted upon by bending movement with no shear force. Therefore the stresses on any cross section must constitute a coule equal in magnitude to the bending moment.

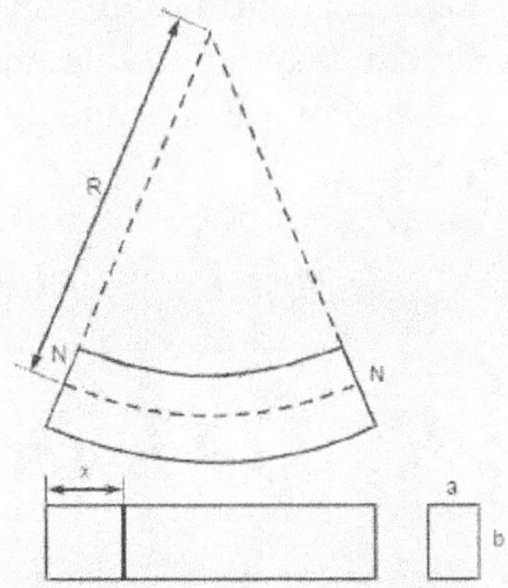

Figure 4.3 Bending of a beam

In order to develop a differential equation of the model, the following assumptions are made.

- The material is homogeneous, isotropic and has the same Young's Modulus (E).
- The beam is initially straight and all longitudinal fibres bend into a circular arc with a common centre of curvature.

- The transverse section remains plane and perpendicular to the neutral surface after bending.
- The radius of curvature (R) is large compared with the dimensions of the cross section.
- The beam is perfectly elastic so that Hooke's law is obeyed.

Considering equilibrium at any cross section of the beam, the internal moment, EI/R due to the structure of beam must e balaced by the external moment (M), i.e.,

$$\frac{EI}{R} = M$$

But,

$$R = \left[\frac{1}{\dfrac{d^2 y}{dx^2}} \right]$$

where,

y is the deflection and

x is the istance from one end of the beam.

Therefore, $EI\dfrac{d^2 y}{dx^2}$ is the desired mathematical model of the system conceived. This example illustrates the need for making assumptions in order to minimize complexity of the situation and the assumptions are context-dependent.

4.5.3 Dynamic Model

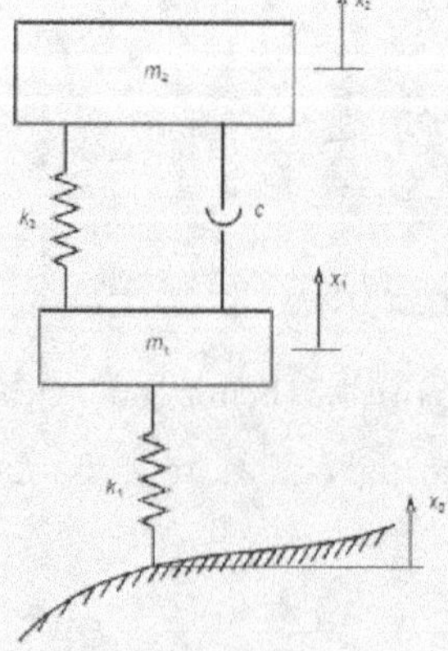

Figure 4.4 Quarter-car model

The system described above is shown in Figure 4.4, a translational model depicting the motion of one wheel of a vehicle (quarter-car model). Here, the stiffness of tyre is modelled by a linear spring (k_1); tyre, axle and other moving parts by a mass (m_1), suspension system by a spring with stiffness (k_2), viscous damper, i.e., shock absorbr as (c) and the support vehicle mass as (m_2).

We can use Newton's second law to develop the equation of motion for each mass and apply equilibrium of forces at the point where the elements join. Displacements x_1 and x_2 are measured from equilibrium position. As the springs are likely to be in compression most of the time, we can assume that spring forces are compressive and represet sprng element forces acordinly.

For mass

$$\sum f_x = m_1 \frac{d^2 x_1}{dt^2} \text{ or } m_1 \ddot{x}_1 \tag{1}$$

i.e., $\quad k_1(x_0 - x_1) - k_2(x_1 - x_2) - c(\dot{x}_1 - \dot{x}_2) = m_1 \ddot{x}_2 \tag{2}$

For mass m_2,

$$\sum f_x = m_2 \frac{d^2 x_2}{dt^2} \text{ or } m_2 \ddot{x}_2 \tag{3}$$

i.e., $\quad k_2(x_1 - x_2) - c(\dot{x}_1 - \dot{x}_2) = m_2 \ddot{x}_2 \tag{4}$

Thus the resulting equations that descibe the motion of two masses are,

$$m_1 \ddot{x}_1 + c(\dot{x}_1 - \dot{x}_2) + k_2(x_1 - x_2) + k_1 x_1 = k_1 x_0(t) \tag{5}$$

These two equations constitute the dynamic mathematical model of the suspension system considered. They constitute the fourth-order system comprising two second-order equations.

4.6 EMPIRICAL MODELS

4.6.1 Model for Expository Analysis

It was stated earlier that any mathematical model developed using experimental data or empirical observations comes under empirical category. Such empirical models are used as descriptive models for explaining the behaviour for better understanding or for predictive purposes. A predictive type usually implies the format and not vice-versa. While in theoretical models several assumptions which are normally verified are employed, empiricism refuses to admit any assumptions that cannot be verified independently by experiment or by analysis of experimental data.

Let us consider an example. A metallurgist wants to analyse yield strength of hot-rolled ferritic–pearlitic steel using different chemical compositions. He conducted several experiments with samples of the specimen having different composition consisting of manganese (Mn), silicon (Si), volume fraction of pearlite and the grain size and obtained yield strength of specimen for each combination. The composition range was 0 to 0.3% by weight of carbon, 0 to 1.67 of manganese, and 0 to 0.87 of silicon. Using the results of the experiment he hasdeveloped a multi-regression equation of the form

$$y = a_0 + a_1x_1 + a_2x_2 + a_3x_3 + a_zx_a$$

With the experimental results, he calibrated the model by fitting the equation tothe data and obtained the vlues of the coefficient a_0, a_1, ..., a_4. The calibrated model is,

$$y = 13.29 + 5.9x_1 + 10.21x_2 + 0.22x_3 + 0.476x_4^{-5}$$

where y represents yield strength in 1000 psi; x_1, x_2, x_3 and x_4 represented manganese, silicon, percentage pearlite and mean linear ferritic intercept grain size respectively. The correlation coefficient for this model is 0.89, which indicated good fitment of experimental data to the model.

The researcher can employ this calibrated model to analyse the sensitivity of yield strength for different compositions of the ingredients considered. Thus the empirical model avoids the necessity for further experiment at additional costs.

For developing this mathematical model using multiple linear regression equations, the researcher should possess adequate theoretical knowledge as to how the various ingredients added, affect the yield strength. The researcher has to ensure that each variable is independent and their contributions are additive in nature. Thus, the domain knowledge is very essential for model development.

4.6.2 Model for Prediction

City planners for preparation of perspective plan for developing transport facilities conduct household survey once in 10 to 15 years to estimate the travel demand. Travel demand prediction is accomplished by a four-step process, viz., trip generation, trip distribution, modal choice and trip assignment. Each step involves development and use of mathematical models appropriate to the context and availability of empirical data. The first step—trip generation—involves determination of number of trips generated by or attracted to each zone of the city in a day. By accumulating the household trips for each zone the total number of trips generated by each zone regardless of destination is obtained.

Let us consider the case of a planner developing a trip generation model for a city. The planner has taken into account various socio-economic factors pertainingto household of each zone and the model considered is

$$T_i = a_0 + a_1 x_{1i} + a_2 x_{2i} + a_3 x_{3i} + a_4 x_{4i}$$

where,

T_i = Number of person trips per household from zone i

x_{1i} = Number of household/acre in zone i

x_{2i} = Number of automobile/household in zone i

x_{3i} = Distance of zone i from Central Business District (CBD)

x_{4i} = Average family income in zone i.

With the help of the empirical data obtained from the household srvey, the model was calibrated and the final model is

$$T_i = 4.33 - 0.005 x_{1i} + 3.39 x_{2i} - 0.1 x_{3i} - 0.128 x_{4i}$$

For developing the model the planner has used knowledge of transportation planning to decide the socio-economic factors that should be considered and incorporated in the model for the given situation, feasibility of data gathering, etc. This model can be used to make prediction of future scenario by estimating the factors considered separately say after a 10 year period, and then incorporating them into the model to get the future trip generation scenario.

4.7 MODEL EVALUATION

A crucial part of the modelling process is the evaluation whether or not a given mathematical model describes a system accurately. Answering this question involves several types of evaluation.

Model evaluation is carried out in two stages, viz., i) model verification, and ii) model validation. Model verification is the process of ensuring that the conceptual description of the model reflects the real world. For instance, if the behaviour of the system of interest is non-linear, then this non-linear behaviour must be reflected in the equations underlying the model. In case a computer is used to manipulate the model, the program should be properly coded and debugged so that every step replicates what we wanted to do. The model predictions have to be verified to ensure its consistency with the data used to construct the model.

Often a mathematical model is developed by abstracting a real system. While developing the model, assumptions are made to simplify the model. It is necessary that the rationale of these assumptions have to be established and verified. Arising out of this, if there are any limitations in the mathematical model, they should also be indicated. For example, if a mathematical model involving elasticity is made, the limitation that it is valid only within the elastic limit has to be specified.

Model validation quantifies uncertainties of a model by comparing head-to-head model predictions to a real-world data. The validation exercise has to be made at several levels like component level, whole system level, etc. The degree of confidence of a model is measured by its validation database. Hence, it is imperative that validation of the model is conducted against experimental data gathered under different operating conditions.

Any model which is not pure white-box contains some parameters that can be used to fit the model to the system it describes. If a model is developed using neural network, the optimization of parameters is made by training. In more conventional modelling through explicitly given mathematical functions, parameters are determined by curve fitting.

An easier way of model validation is to check whether a model fits experimental measurements or other empirical data. In models with parameters, a common approach is to split the data into two sets—one set for training and the other for verification. The training data are used to estimate the model parameters. Any accurate model will closely match the verification data even though this set of data was not used to estimate the parameters. This practice is known as cross-validation. Defining a metric to measure the deviation between observed and projected data is a useful tool of assessing a model fit.

It is difficult to test the validity of the general mathematical form of a model. Mathematical tools are available to test the fit of statistical models than models involving differential equations. Tools from non-parametric statistics can sometimes be used to evaluate how well data fits to a known distribution.

Assessing the scope of the model, that is, to determine situations in which the model is applicable can be less straightforward. If the model was constructed based on a set of data, one must indicate the aptness of the dataset vis-a-vis the system or phenomenon modelled. The model should also describe well the properties of the system between data points, which is called "interpolation". If it is capable of describing events or data points outside the range of observed data, it is called "extrapolation". As the purpose of modelling is to increase our understanding of the situation/ phenomena of real world, the validity of the model rests not only on its fit to empirical observations but also on its ability to extrapolate situations or data beyond those originally described in the model.

Technique of Dimensional Analysis (DA) is used to develop theoretical models to explain and/or predict the behaviour of a phenomenon, especially when the relationship between dependent and independent variables is not clearly known. A model developed in this case will be dimensionless. This theoretical model developed using DA can be used to cross check the empirical model developed (say a multi-regression model) based on experimental data of the researcher. Such cross validation helps to ensure the reliability of empirical model developed.

4.8 LIMITATIONS OF MATHEMATICAL MODELS

Thousands of mathematical models have been developed successfully covering wide range of disciplines in order to gain insights and/or for predictions covering a variety of situations. Nevertheless, there are still equally large number of situations which have not yet been mathematically modelled either because the situations are sufficiently complex or because the models developed are mathematically intractable.

Advancements in computer technology have enabled more number of situations to be mathematically modelled. Not only more models were developed but also more realistic models to obtain better agreement with observations have been developed. Successful guidelines are necessary for choosing the number of decision parameters and to estimate the values of these. It will be very useful to modellers. Reasonably accurate models can be developed involving 5 or 6 decision parameters. But if more number of parameters are involved, the model's complexity gets increased. An analyst generally aspires to involve minimal number of decision parameters and also wants to estimate them more accurately.

When models become too complex and mathematically intractable, the analyst has no other option but to resort to numerical solution using simulation approach. This is a form of empirical study and is the last resort when problems are highly complex, ivolving too many decision parameters and probabilistic situations.

5
Simulation and Dimensional Analysis

5.1 IMPORTANCE OF SIMULATION

Simulation refers to the application of computational models to the study and prediction of physical events or the behaviour of engineered systems. The development of computer simulation has drawn from a deep pool of scientific, mathematical, computational and emerging knowledge and methodologies. With the depth of its intellectual development and its wide range of applications, computer simulation has emerged as a powerful tool, one that promises to revolutionize the way engineering and science are conducted in the 21st century.

Computer simulation represents an extension of theoretical science, in that it is based on mathematical models. Such models attempt to characterize the physical predictions or consequences of scientific theories. Simulation can be much more. For example, it can be used to explore new theories and to design new experiments to test these theories. Simulation also provides a powerful alternative to the techniques of experimental science and observation when phenomena are not observable or when taking measurements are impracticable or too expensive.

Although the use of computer simulation in engineering science began over half a century ago, only in the past decade or so have simulation theory and technology made a dramatic impact across the engineering fields. That remarkable change has come about mainly because of developments in the computational and computer sciences and the rapid advancement in computing equipment and systems.

Simulation-Based Engineering Science (SBES) is considered as a discipline that provides the scientific and mathematical basis for the simulation of engineered system. Such systems range from microelectronic devices to automobiles, aircraft, and even infrastructures of oil fields and cities. SBES fuses the knowledge and techniques of the traditional engineering fields—electrical, civil, mechanical, chemical, aerospace, biomedical, nuclear and material science—with that of fields like computer science, mathematics and the physical and social sciences. As a result, engineers are better able to predict and optimize systems affecting almost all aspects of our lives and work, including our environment, security and safety and the products we use.

A panel appointed by the USA Government in 2004 observed that SBES is an indispensable discipline to their continued leadership in science and engineering. They have further added that it is central to advances in biomedicine, manufacturing, security of the homeland, micro-electronics, energy and environmental sciences, advanced materials and product development. This new discipline could significantly impact every aspect of human experience.

5.2 SYSTEM SIMULATION

System simulation consists of a mathematical model of a real world situation called "the system", in such a way that the model behaves like the actual system to events and inputs that take place over time. It is the process of designing a model of a real system and conducting experiments with this model for the purpose of understanding the behaviour of the system or evaluating various strategies for the operation of the system.

By collecting data on the system's response under various conditions, it is possible to learn how the real system may perform, without having to attempt costly experiments with the actual system. The main value of simulation is that the simulation model can be conveniently manipulated until the designed results are obtained. Simulation can be performed on existing systems to identify options for problem-solving or system improvements. It is also an excellent forward planning tool to evaluate and compare the proposed system before they are acquired and installed.

Figure 5.1 indicates the different ways in which a system might be studied and the position of simulation in such a stdy.

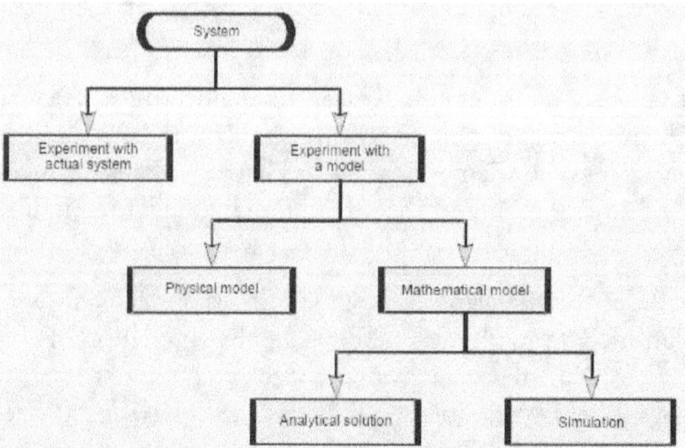

Figure 5.1 Different ways of studying a system

5.3 CLASSIFICATION OF SIMULATION MODELS

We have a mathematical model to be studied by means of simulation and then we must look for a particular tool to do this. It is useful for this purpose to classify simulation models along three different dimensions.

- *Static vs dynamic simulation models* A static simulation model is a representation of a system at a particular point in time. A dynamic simulation model represents a system as it evolves over time such as a conveyor system in a factory.

- *Deterministic vs stochastic* If a simulation study does not contain any probabilistic (i.e., random) components, it is called deterministic (e.g., a system of differential equation describing a chemical reaction). If a system modelled has at least one random component (input), this gives rise to stochastic simulation.

- *Continuous vs discrete simulation* Discrete event simulation concerns the modelling of a system as it evolves over time by a representation in which the state variables change instantaneously at separate points in time. Continuous simulation concerns the modelling over time of a system by a representation in which the state variable changes continuously with respect to time.

5.4 STEPS IN A SIMULATION STUDY

There are various steps involved in a discrete event simulation study. Figure 5.2 shows the steps that will compose a typical simulation study. It should be noted that a simulation study is not a simple sequential process. As one proceeding with the study, it may be necessary to go back to a previous step.

Even though the figure is self-explanatory, certain steps may need some explanation. For instance, to check conceptual model validity, we have to perform a structured walk-through of the conceptual model to check the rationale of the assumptions to ensure that the model's assumptions are correct and complete. The model may be programmed in any programming language (e.g., C or FORTRAN) or using simulation software (e.g., Arena, Automod, Extend, MATLAB, etc.). The use of simulation software reduces not only the programming time, but also the cost of the project.

To check the validity of the programmed model, if there is an existing system, the performance of the model and existing system is compared. The model results should be reviewed for correctness. Sensitivity analysis should be used to determine what model factors have significant impact on performance measures and thus, have to be modelled carefully. In design of experiments, the length of each run, length of warm-up period, and number of independent simulation runs using different random numbers have to be taken ino account.

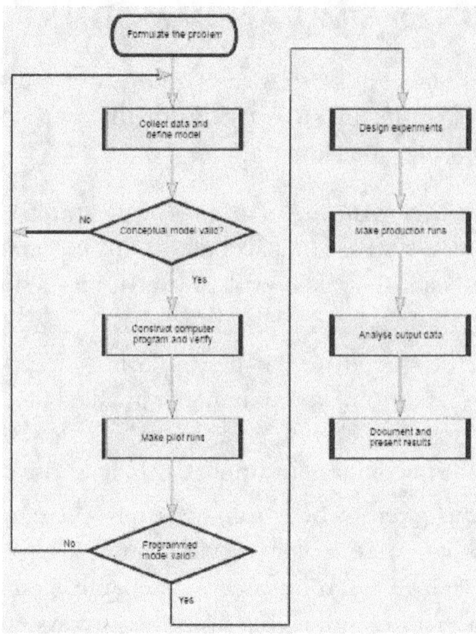

Figure 5.2 Steps in a typical simulation study

5.5 SIMULATION SOFTWARE

The simulation packages are classified into two major types, viz., simulation languages and application-oriented simulators. Simulation languages are general in nature, and the model is developed by the user writing the code. Simulation languages, in general, provide a great deal of modelling flexibility, but are often difficult to use as it consumes time. On the other hand simulators are oriented towards a particular application, and a model can be developed by using graphics, dialog boxes and pull-down menus. Simulators are easier to learn, but may not be flexible enough for some problems. In recent years vendors of simulation languages are attempting to make their software easier to use by employing a graphical model-building approach. Similarly, the vendors of simulators make their software more flexible by allowing programming in certain model locations using an internal pseudo-language.

A general-purpose simulation package can be used for any applications, but might have special features for certain functional applications (e.g., for manufacturing, communications). On the other hand, application-oriented simulation package is designed to be used for a certain class of applications. Examples of some application-oriented simulation packages include; Manufacturing–Automod, Extend, Pro Model, QUEST, Arena; Communication Networks–COMNET, OPTNET; Process Engineering–Process Model, SIMPROCESS; Health care–Med Model.

5.6 VALIDATION AND DATA COLLECTION

5.6.1 Validation

The consistency of a simulation model built by a researcher with the real system as it existed before any changes were made has to be checked before the researcher can claim that the simulation model is a useful tool for studying the behaviour under new hypothetical conditions. By validation we mean a study as to how well the behaviour of a model replicates that of the true system. Validation is an important step in any simulation experiment and a simulation model which has not been validated cannot command any credibility and the whole effort of the experimenter is nullified.

Validation and analysis of simulation study is a continuous process that begins from the start of the experimentation. Confidence is built into the model as we proceed with the study. Several authors have put forth several practical considerations for validating simulation results. Van Horn proposed a three-phase multistage validation process:

    i.   Constructing a set of hypotheses and postulates for the process using all available information—observations, general knowledge, relevant theory and intuition.

    ii.   Attempting to verify the assumptions of the model by subjecting them to empirical testing.

    iii.   Comparing the input–output transformations generated by the model to those generated by the real world.

Fishman and Kiviat have segmented the process of evaluation of the simulation model into three categories, viz.,

    i.   Verification to ensure that the actual model behaves the way the experimenter wants it to.

    ii.   Validation to test the agreement between the behaviour of the model and that of the real-world system.

    iii.   Problem analysis which deals with the analysis and interpretation of the data generated by the experiments. This approach requires the analyst to first build confidence in the logical structure of the model and then establish the credibility of the model's behaviour by comparison with the real system abstracted.

In an era where large number of dedicated application-oriented simulation software are used by the analyst, verification and validation assumes considerable significance. Many models are built using constructs (standard modules) and the internal structure of these modules may not be fully known to the analyst. Therefore, before using any particular software, he/she has to extensively test it using known examples to have a proper understanding of the software. This will ensure that the model built by the analyst using the software will be realistic.

Many statistical methods such as analysis of variance, regression analysis, chi-square test, spectral analysis, etc. are used to compare that model output with the real-world data.

5.6.2 Data Collection

Simulation is a statistical experiment. Its output data must be interpreted using proper statistical inference tools such as confidence interval and hypothesis testing. The data are supposed to be generated out of a distribution (stationarity) and should not be correlated (independent) and comes from a normal population. But however, on many situations, these cannot be fully satisfied, yet we assume that the observations satisfy the conditions of stationarity, independence and come from a normal population.

Simulation output is a function of the length of a simulated period. There will be transient effect during the warm-up period before the system reaches a steady-state condition. The concept of transient and steady states are relevant only in respect of a non-terminating simulation, i.e., the simulation that applies to systems whose operation continues indefinitely. In the case of terminating simulation wherein an operation starts at a particular time and then closes also at a specified time repeatedly, the transient behaviour is a part of the normal operation of the system and hence cannot be ignored.

Three common methods are used to collect data from simulation study. They include,

i. Sub-interval method

ii. Replication method

iii. Regenerative method

The sub-interval method calls for truncating the initial transient period first and then subdividing the remainder of the simulation run into n equal sub-intervals. The average of each sub-interval is used to represent a single observation (data). The advantage of this method is that the effect of transient conditions is mitigated. The disadvantage of this approach is that successive batches are correlated due to the common boundary.

In the replication method, each observation is represented by an independent simulation run in which the transient period is truncated. The computation of averages for each batch is the same as in the sub-interval method. The only difference is that the standard variance formula is applicable because the batches are not correlated.

The regeneration method may be considered as an extended case of the sub-interval method. This method attempts to reduce the effect of autocorrelations because we adopt similar starting conditions for each batch. However, unlike the sub-interval method, the nature of the regeneration method may result in unequal time bases for the different batches.

5.7 MAJOR APPLICATIONS

The major application of simulation in some of the areas is given below. These examples are only illustrative and not exhaustive.

Infrastructure

- Simulation-based studies can help detect and measure the presence of biological or chemical contaminants in the air, and given the weather data, identify the likely release location and magnitude of the release. This will help to evolve an optimal response plan.

- Optimization of designs of buildings and other infrastructure elements. Such designs could be site-specific, could interact well with natural and manmade surroundings and could blend with other urban systems of which they are a part.

- Can predict the effects of effluents from existing and proposed facilities in urban and natural environments. This will increase the reliability and usefulness of environmental impact studies.

- Can be used to assist in the design and placement of air and water contaminant disposal and flood abatement.

Energy and environment

- Energy-related industries rely on modern simulation methods to monitor the production of oil reservoirs, plan remedial measures for pollution, and devise control strategies.

Material science

- Multi-scale modelling and simulation are transforming the science and technology of new material development and improvement of existing materials. The new methods enable unprecedented ability to manipulate metallic, ceramic, semiconductor and polymeric materials.

- The principle of materials design is rooted in the correlation of molecular structure with physical properties. From these correlations, models can be formulated that predict micro-structural evolutions. Such models allow the researcher to investigate the mechanisms underlying the critical behaviours of materials and to systematically arrive at improved designs.

Industry

- Plays a significant role in the design of materials, manufacturing processes and products. Increasingly, simulation is replacing physical tests to ensure product reliability and quality.

- In automotive industry crash-worthiness studies are made using simulation instead of using real vehicles and dummies.

- Simulation is used in petrochemical plants to develop steady-state process and to perform detailed optimization. As a result, chemical plants are more energy-efficient and environment-friendly.

- Over the past two decades the integrated circuit industry has been a major player in simulation-based engineering. Highly integrated easy-to-use software such as SPICE is used for circuit analysis. With the clock rates moving into gigahertz range, circuit theory may not be of much use. Future-generation transistors such as single-electron transistors, low-threshold transistors and quantum computing devices will be based on new physics that links quantum mechanics and electromagnetics.

5.8 DIMENSIONAL ANALYSIS

5.8.1 Importance of Dimensional Analysis

Dimensional analysis (DA) is an important and useful technique to understand the properties of physical quantities independent of the units used to measure them. It is extensively used for the investigation of problems in many branches of engineering and science. This technique was developed by the 19th century French mathematician Joseph Fourier based on the idea that physical laws like Force = mass × acceleration ($F = ma$) should be independent of units employed for the measurement of physical variables.

- Dimensional analysis has several advantages to the user and they include the following.

- Dimensional analysis (DA) is a form of symbolic modelling which is of immense use in modelling and experimentation, especially when the functional relationship between the factors considered are not well-established.

- It helps the investigation by reducing the number of factors (variables) by grouping them in the problem, which in turn reduces the labour of experimental investigation.

- The analyst need not vary each individual factor, instead he/she can vary each dimensional group as a whole to study the performance.

- Dimensionless groups are useful means of defining the conditions that exist in a physical system and indicating which properties are significant.

- The number of curves that need to be plotted and referred to in the analysis or design calculations are greatly reduced.

- Dimensional analysis is generally used to check the validity of derived equations and computations.

- It helps to formulate reasonable hypothesis about complex physical phenomena or situations, which in turn can be tested by experimental methods.

- It helps to arrive at qualitative solutions to problems by dimensional reasoning followed by experimental investigation thereby enabling complete solution to the problem.

Dimensional analysis is thus an important tool, which a researcher or analyst has to be conversant with. However, only an introduction is given here supplemented by a case study.

5.8.2 Dimensional Quantities

A physical quantity is a conceptual property of a physical system, and it can be expressed numerically in terms of one or more standards. The dimension is the relationship of a derived quantity to whatever primary quantities have been selected. Most of the physical quantities are expressed in terms of a combination of five basic dimensions; mass (M), length (L), time θ(T), electric charge (Q) and temperature (). These dimensions are considered as basic because they are easy to measure in experiments. Dimensions are not the same as units. For example, the physical quantity speed (velocity) is measured in terms of kilometer per hour or metre per second, etc. Nevertheless, the speed is always calculated as length (L) divided by time (T). Therefore, the dimension of speed or velocity is $(L)/(T)=(LT^{-1})$.

Like speed, there are many such dimensional quantities which are combination of fundamental units. Some of them are,

Area = Length × Length = $(L)(L) = L^2$.

Acceleration = Velocity/Time = $(LT^{-1})/(T) = LT^{-2}$.

Force = Mass × Acceleration = $(M)(LT^{-2}) = MLT^{-2}$.

Pressure = Force/Area = $(MLT^{-2})/(L^2) = ML^{-1}T^{-2}$.

While there are many physical quantities with dimension, some are dimensionless. For instance, 'strain', is denoted by elongation divided by original length. This implies length divided by length $(L)/(L) = L^1 L^{-1} = L^0 = 1$, a pure number and hence dimensionless. Similarly, trigonometric functions such as $\sin\theta$, and exponential functions are also dimensionless. Frequency is another quantity given by cycles per second, and it is dimensionless though second is a time unit and cycles are dimensionless (number).

5.8.3 Development of Dimensional Relations

Principle of dimensional homogeneity   While analysing physical phenomena, one tries to connect the response factor (output) with the factors that are responsible for causing the response (input) by establishing appropriate equations. In algebraic equations, the terms that are added or subtracted must have the same dimensions. This implies that each term on the left-hand side of an equation must have the same dimensions as each term on the right-hand side. An equation in which each term has the same dimension is said to be dimensionally correct. This leads to the principle of homogeneity, which states that if an equation truly expresses proper relationship between factors in a physical process, it will be dimensionally homogeneous.

In an equation there could be dimensional factors, dimensional constants. Consider the equation, $v = u_0 + at$, where u_0 = the initial velocity, a = constant acceleration and v = the final velocity. The factors v and t vary during a given case and can be reckoned as dimensional factors (dimensional variables), while u_0 and a are dimensional constants as they are held constant during the duration of the given case. In addition, there can be pure constants which are simply numbers in a given case.

Many methods are there for performing dimensional analysis which include, indical method and method of Buckingham π theorem. The Buckingham π theorem approach offers a generalized method for obtaining solution.

Buckingham π theorem   There are two theorems. According to the first one, if there are m factors representing physical properties such as velocity, density, etc., the relationship between them can be expressed by means of (m – n) dimensional groups of factors called π groups, where n is the number of dimensions such as mass, length, time, etc., used to express them. Suppose a physical law can be epressed as, $\phi(f_1, f_2, ..., f_m) = 0$ Then as per the first theorem, $\phi(\pi_1, \pi_2, ..., \pi_{m-n}) = 0$.

The second theorem stipulates that each π group is a function of n governing or repeating factors plus one of the remaining factors. Repeating factors, refer to the factors that appear in most of the π groups and influence the problem.

Aspects to be considered   The following aspects may be of use to an analyst while performing dimensional analysis.

i.   First identify all the independent factors to be considered in the problem. It is better to err on the safe side in this respect.

ii.   Prepare the dimensional matrix after checking the dimensions of each factor.

iii.   Identify the repeating factors. A repeating factor generally includes all the dimensions involved to describe the phenomena. It is not a necessary condition and due consideration has to be given for practicality.

iv.   For each dimensional group asign a π term. There will be (m – n) number of π terms with m as total number of factors and n as he number of repeating factors.

v.   For each π term develop an equation linking the associated repeating factors raised to an appropriate exponnt and ne ofthe non-repeating factor (eg., $\pi = (x_1)^a (x_2)^b (x_3)^c x_4$ where and are repeating factors while is a non-repeating one).

vi.   Substitute dimension for each factor and evaluate the values of the exponent and then formulate the equation.

Dimensional analysis uses both the principle of dimensional homogeneity and Buckingham π theorem to solve many real-life problems wherein physical phenomena are involved. In the next section we deal with a case study to demonstrate how dimensional analysis was carried out by a researcher using Buckingham π theorem.

Case Study

Electricity distribution system generally uses porcelain insulators in outdoor electrical power transmission lines. A researcher tried to develop a new polymeric material for insulators with a view to enhance properties and reduce the maintenance. To evaluate the newly designed polymeric housing material for the insulator, he considered the following factors, viz., i) tracking resistance (TR), ii) arc resistance (AR), iii) dielectric strength (DES), iv) dielectric current (DC), v) dissipation factor (DF), vi) volume resistivity (VR), vii) surface resistivity (SR), viii) tensile strength (TS), ix) percentage elongation (PE), and x) comparative tracking index (CTI). His objective was to establish a theoretical model (relationships) that relate tracking resistance with other electrical properties, and similarly the tensile strength with other electrical properties using dimensional analysis technique.

Out of the ten factors considered, dielectric constant (DC), dissipation factor (DF), and percentage elongation (PE) are dimensionless. Hence, the remaining seven factors are considered for the analysis. As only four fundamental dimensions M, L, T and Q are used, it became necessary to identify four repeated variables to express the relationships. As he was interested in developing equations for tracking resistance (TR) and tensile strength (TS) only, he decided to club the factor volume resistivity (VR) and arc resistivity (AR) by introducing a new factor as (VR/AR).

The dimensions of chosen factors expressed in M, L, T, Q are,

TR	:	$[T]$	DES	:	$[MLT^{-2}Q^{-1}]$
TS	:	$[ML^{-1}T^{-2}]$	VR	:	$[ML^{3}T^{-1}Q^{-2}]$
AR	:	$[T]$	SR	:	$[ML^{2}T^{-2}Q^{-1}]$

The dimensional matrix of the six factors nd the four fundamental dimensions is as follows:

	TR	TS	VR/AR	DES	SR	CTI
M	0	1	1	1	1	1
L	0	-1	3	1	2	2
T	1	-2	-2	-2	-1	-2
Q	0	0	-2	-1	-2	-1

As the rank (n) of the dimensional matrix is 4 and the number of factors (m) is 6 there will be two π groups (i.e., $6 - 4 = 2$) π_1 and π_2 respectively, to express the relationship of TR and TS, the repeating factors.

Following the Buckingham, π theorem, these tw relatonships were established for TR and TS as,

$$\pi_1 = (CTI)^{a_1} (DES)^{b_1} (SR)^{c_1} (VR / AR)^{d_1} TR \tag{1}$$

$$\pi_2 = (CTI)^{a_2} (DES)^{b_2} (SR)^{c_2} (VR / AR)^{d_2} TS \tag{2}$$

Expressing in terms of dimension and introducing constants D_1 and D_2 for the unacconted three dimensionles factors.

For equation 1,

$$M^0 L^0 T^0 Q^0 = D_1 \left[ML^2 T^{-2} Q^{-1} \right]^{a_1} \left[MLT^{-2} Q^{-1} \right]^{b_1} \left[ML^2 T^{-1} Q^{-2} \right]^{c_1} \left[ML^3 T^{-2} Q^{-2} \right]^{d_1} [T]^1 \tag{3}$$

For equation 2,

$$M^0 L^0 T^0 Q^0 = D_2 \left[ML^2 T^{-2} Q^{-1} \right]^{a_2} \left[MLT^{-2} Q^{-1} \right]^{b_2} \left[ML^2 T^{-1} Q^{-2} \right]^{c_2} \left[ML^3 T^{-2} Q^{-2} \right]^{d_2} \left[ML^{-1} T^{-2} \right] \tag{4}$$

Equating the powers of equation 3 n both sides the following equations are obtained.

$$a_1 + b_1 + c_1 + d_1 = 0$$

$$2a_2 + b_2 + 2c_2 + 3d_2 = 1$$

$$2a_2 + 2b_2 + c2 + 2d_2 = -2$$

$$a_2 + b_2 + 2c_2 + 2d_2 = 0$$

By solving these equations, the values of exponents are $a_1 = 1, b_1 = -1, c_1 = 1, d_1 = -1$.

Substituting the values of exponents in equation 3, the following equations for the tracking resistance (TR) were obtained.

$$TR = D_1 [CTI]^1 [DES]^{-1} [SR]^1 [VR/AR]^{-1} \tag{5}$$

Replicating the above steps for equation 4, he following simultaneous equations were obtained.

$$a_2 + b_2 + c_2 + d_2 = -1$$

$$2a_2 + b_2 + 2c_2 + 3d_2 = 1$$

$$2a_2 + 2b_2 + c2 + 2d_2 = -2$$

$$a_2 + b_2 + 2c_2 + 2d_2 = 0$$

By solving the above equations, the values of exponents are $a_2 = 0$, $b_2 = -2$, $c_2 = 0$ and $d_2 = -1$.

Introducing the values of exponents in equation 4, the relationship obtained for tensile strength (TS) is,

i.e., $\quad\quad TS = D_2 \, [CTI]^0 \, [DES]^2 \, [SR]^0 \, [VR/AR]^{-1}$

$$TS = D_2 \, [DES]^2 \, [AR/VR]^1$$

This demonstrates how the technique of dimensional analysis helped the researcher to establish relationships and how further experimental works have to be carried out to establish the values of constants, D_1 and D_2.

5.9 SIMILITUDE

5.9.1 Meaning and use of Similitude

The word "similitude" connotes similar or comparison. Often both similitude and similarity are used synonymously. The concept of similitude is used in the testing of engineering models, especially in hydraulics, automobile and aerospace engineering.

As testing of the prototype of a proposed design for evaluating its performance is not only time-consuming but also costly, engineers and scientists often use scale models to evaluate. These scale models are usually smaller than the final design. For example wind tunnel tests on scale models of aircraft, automobiles, trains and submarines are carried out to study the aerodynamic characteristics. In civil engineering models of rivers, harbours, and dams are employed to study the tidal flow conditions. Even study of the airflow around buildings can be studied with the help of models. In fact testing of models is a critical step in the design and development of a product.

The results of model testing greatly depends upon the extent to which a model is to be an exact replica of the prototype, i.e., similitude. For a model is said to have similitude with the real application (prototype), certain similarity conditions are to be satisfied.

Using the technique of dimensional analysis, the analyst decides the grouping of factors related to the phenomena under the given set of conditions. In fact these factor groupings are of much help in the model study. The analyst instead of varying each factor in the group, can vary each group as a whole systematically to determine their relative importance. This type of evaluation is extensively used in fluid flow-related problems.

Several formulae and dimensionless quantities have been developed for many engineering problems and they are reported in published literature for easy reference by the users. By using similitude, it is possible to predict the performance of a new design based on the data from the existing design. Further, similitude and models are extensively used in the validation of computer simulations and this eventually may eliminate the need for physical models.

5.9.2 Types of Similitude

The scale models are often smaller than final prototype and when analyses are made, one has to determine beforehand under what conditions the model is to be tested. While the shape may be the same, when the size is scaled down, other parameters such as pressure, temperature, velocities and fluid type need to be altered. Similitude is achieved only when the analyst ensures that the test results are applicable to the prototype. This implies that both the model and the prototype meet certain criteria. They include i) geometric similarity, ii) kinematic similarity and iii) dynamic similarity.

Geometric similarity   This is the first type of similarity a model must satisfy, however, it may be difficult to comply while modelling rivers, etc. Essential conditions include identity of shape, equality of corresponding angles or arcs and constant proportionality or scale factor corresponding to linear dimensions. The scale factor is defined as model dimension/prototype dimension, i.e., $s = L_m/L_p$ where s is the scale factor, L_m refers to model dimension and L_p refers to prototype dimension. For example, if an analyst builds a model of an automobile to study aerodynamic effects in a wind tunnel, he has to scale down length, breadth and height proportionate to the scale factor, but must retain the shape of all corners as that of the prototype.

Kinematic similarity   Kinematics deals with motion of objects without reference to the forces that cause the motion. As far as kinematic similitude is concerned, it is based on the ratio of the time proportionality between corresponding events in the prototype and the model developed. Thus, it represents the time-scale ratio. Frequently, it is combined with the length dimension to express ratios such as velocity ratio, volume flow rate ratio, etc., at corresponding positions. The velocity-scale ratio will be the same and kinematiclly similar for both the model and the prototype.

For example,

$$\text{Ratio of velocities } (V_r) = \frac{\ddot{u}\ddot{u}\ddot{u}\ddot{u}\ddot{u}\ddot{u}}{\text{Pr ototype velocity}}\frac{V_m}{V_p}$$

$$= \frac{(L_m)(T_m^{-1})}{(L_p)(T_m^{-1})} = (L_r)(T_r^{-1})$$

where, L_r = Length ratio and T_r = Time ratio.

Dynamic similarity The dynamic similarity requires that the forces acting on the corresponding locations both in the model and the prototype should bear a fixed ratio. In the case of fluid flow, the forces may be caused due to inertia, viscosity, gravity, pressure, vibration, surface tension, etc. Dynamic similarity automatically ensures kinematic similarity in certain situations where the gravitational forces cause the fluid motion. It is also possible to obtain kinematic similarity without ensuring dynamic similarity in cases where the movement is caused mechanically.

It is normal practice in dimensional analysis to form dimensionless groups which are useful to the analyst especially in respect f flow analysis. One such group is the familiar Reynolds number (R_n). It is the ratio of inertial forces and viscous forces and defined as, $R_n = \rho V L \big/ \mu = \dfrac{LV}{v}$ where $v = \mu/\rho$ is the kinematic viscosity.

Reynolds number must be the same for both the mdel and the prototype when flow studies are conducted. Therefore,

$$R_n = \frac{L_m V_m}{v_m} = \frac{L_p . V_p}{v_p}$$

$$V_m = \left(\frac{L_p}{L_m}\right)\left(\frac{v_m}{v_p}\right)V_p = \frac{1}{s}\frac{v_m}{v_p}V_p$$

where s is the scale factor. If we build a 1 to 20 scale model (s = 0.05), the velocity in the model has to be increased by a factor of 20 to maintain dynamic similarity.

There are other similarities such as thermal similarity and chemical similarity. Thermal similarity requires that the temperature profile both in the model and prototype must be geometrically similar at corresponding times. In the case of chemical similarity, it is obligatory that the rate of chemical reaction at any location in the model be proportionate to the rate of the same reaction at the same location andat the corresponding time in the prototype or the system studied.

6
Probability and Distributions

6.1 IMPORTANCE OF STATISTICS TO RESEARCHERS

6.1.1 Statistics in Engineering

Statistics is concerned with scientific methods for collecting, organizing, summarizing, presenting and analysing data, as well as for drawing valid conclusions and making reasonable decisions on the basis of such analysis. It is neither possible nor economical for the researcher to perform experiment on all possible ways or gather data about the entire population. Therefore, one has to necessarily resort to sample data from the population. In sampling, probability plays a dominant role. Probability denotes the extent to which an event is likely to occur measured by a number between 0 and 1. Engineers are finding increasingly the practical usefulness of some of the basic concepts of probability and statistics, which is a branch of mathematics. While this chapter is devoted to engineering experimentation involved in research, the statistical concepts and methods discussed are also applicable to other areas. In order to give some practical exposure of statistical tools, the approach adopted here is to concentrate on general concepts and specific calculation methods for commonly occurring problems without emphasis on proofs.

Suppose a researcher buys a carbon steel rod and the manufacturer says that modulus of elasticity (E) of the steel is 206.8 GPa, but while actual testing of specimens of the rod, the E value is found to vary between 205.7 GPa to 207.9 GPa. Here, statistics will be helpful in deciding the probability of the E value falling below a chosen value. Similarly when specimens of 25 mm diameter are prepared using machining, in actual practice one can notice that the variation between specimens could be from 24.5 mm to 25.5 mm. This variation will produce an associated uncertainty in results.

In engineering experimental analysis, one can find that statistical methods have important applications such as specifying the uncertainty of measured data, planning of experiments to get the maximum amount of significant data with the least effort, time and expense. Moreover, the hypothesis developed could be tested rationally to decide the probability of their being true or false.

Every measurement we make may have some inaccuracy and this can be broken down into so-called systemic and random or chance errors. Systemic errors repeat themselves if the measurement is repeated and could be removed by calibrating the instrument and the experimental set-up. Random errors do not behave this way, rather they exhibit scatter. The best that we can do is to statistically estimate what their largest values might reasonably be. We can then quote the measurement as, say $25.20 \pm .06$ to indicate that we are, say 95% sure that the value is somewhere between 25.14 and 25.26.

Many statistical methods were originally developed for use in biological and social sciences, where the interrelationship between variables are not clear-cut as they often are in engineering. On several occasions engineers and physicists have to deal with poorly defined situations or forced-to-use data that have low precision. Proper application of statistical analysis can greatly help them in such situations.

6.1.2 Scope of Statistical Analysis

Often physicists and engineers run experiments involving several variables, but their variables are subject to much more precise control than those in biological and social sciences. Statistics have been found very useful, indeed necessary, in sorting out the subtle effects of various interacting variables in situations of this sort. As engineering is becoming more involved with biological and social system, the need for engineers to have some statistical background increases. Even in pure technical problems, statistics has been found to be very helpful in giving a rational basis for reaching judgments when the interaction of variables is subtle.

There are several uses of statistical tools. Some of the important applications of statistical tools in engineering include the following.

- At the most basic level, we are interested in the analysis of experimental data so that the results can be unambiguously described by appropriate statistical parameters.

- In statistical inference, we use statistics for making reliable decisions utilizing tests of hypothesis and confidence limits. Hypothesis tests are used to determine whether there is a significant difference between the characteristics of an observed set of data and a proposed mathematical model of the data.

- Confidence limits allow us to determine the range in which the true characteristic of a population is likely to lie.

- Analysis of variance is a test for the equality of means and/or variances of group of observations, such as the means resulting from different experimental procedures.

- The statistical design of experiments helps the researcher to collect data in a more efficient and economic way that gives value of information per experiment optimally.

- Regression analysis is another tool used in experimentation to determine the relation between two or more variables and this can also be used for prediction purposes.

6.2 PROBABILITY CONCEPTS

6.2.1 Probability Definition

Probability is a number between 0 and 1 related to a given event. If the event is absolutely certain, its probability is 1 and if the event is impossible, its probability is zero. Three different definitions of probability are in common use, each one is appropriate for certain types of applications.

Classical definition   If an event can occur in N equally likely and mutually exclusive ways, and if n of these ways have an attribute A, then the probability $P(A)$ of the occurrence of A is defined to be n/N. This definition is applicable to study of games of chance. For example, if there are 4 aces in a deck of 52 cards, the probability of drawing an ace in one try is $4/52 = 0.777$.

Frequency definition   If an experiment is carried out N times and an event A occurs n times, then if we let N approach infinity, the limit of n/N is defined to be the probability $P(A)$ of an event A. This is the most popular definition and allows treatment of many practical problems in which classical definition could not be applied.

For example, it is neither desirable nor economical to test all the finished components in a continuous production system. Theoretically, it is not possible to study an infinite population; hence a sample size large enough to be reliable but small enough to be economical is chosen to make decision about the quality of components produced.

Subjective definition   In this the probability $P(A)$ of a proposition A is a measure of the "degree of belief" one holds in the proposition. This is perhaps the broadest definition and a necessary one. Suppose we are trying to choose between two alternative strategies, each of which is likely to produce certain results, and we cannot experimentally try out each case, then one has to rely on "expert judgment" to assign numerical probability in such cases. (e.g., betting in horse race, whether to drop a bomb or not on enemy target). Here again, the classical and frequency interpretation are useless and a subjective judgment is necessary.

Irrespective of the interpretation one puts on probability, once a number has been chosen, any further calculations are independent of the definition.

6.2.2 Probability Laws

Probability theory has special and precise system of language and notation. Two events A and B are said to be 'independent' if the occurrence of one event (say A) has no effect on the probability of occurrence of the other (say B). The two events A and B are called "mutually exclusive" if one of them happens, the other cannot, i.e., they have no elements in common. The various probability laws are,

Multiplicative law

i.   If A, B, C ... are independent events, then the probability that all events occur simultaneously is called joint probability, i.e., the product of their respective prbabilities, $P(ABC) = P(A) \cdot P(B) \cdot P(C)$

ii. If A and B are mutually excusive then, $P(AB) = 0$

iii. If two events A and B are not independnt, then,

$$P \ (A \text{ and } B) = P(AB) = P(A) . P\left(\frac{B}{A}\right) = P(B) . P\left(\frac{A}{B}\right)$$

where,

P(B/A) is the conditional probability of B with respect to A, i.e., probability of B occurring, assuming that A has occurred.

P(A/B) is the conditional probability of A with respect to B, i.e., probability of A occurring, assuming that B has occurred.

Whether events are independent or not is not always obvious or clear-cut in practical applications. By critically examining the situation, one has to decide whether the events are dependent or independent. When the correct interpretation is not obvious after brief consideration, more theoretical and/or experimental studies may be necessary for reaching a decision.

Additive law

i.   The probability that event A or B occur is given by te relation,

$$P \ (A \text{ and } / \text{ or } B) = P \ (A) + P \ (B) - P \ (AB)$$

The events A and B need not be independent, so long as we know their joint probability $P \ (AB)$, i.e., not mutually exclusive events.

ii. If A and B are mutually exclusive (i.e., if one of them happens, the other cannot), then $P \ (AB) = 0$. Inthat case,

$$P \ (A + B) = P \ (A \text{ and } / \text{ or } B)$$

$$P \ (A + B) = P \ (A) + P \ (B)$$

This can be generalized to any number of events by a process of continued eplication,

Baye's rule Baye's method allows us to modify a probability estimate as additional information becomes available. If B is an event, which may be possible by k mutually exclusive and exhaustive ways of A_i, i = 1, 2...k, then

This is known as posterior probability.

6.3 PROBABILITY DISTRIBUTIONS

6.3.1 Errors and Samples

The act of making any type of experimental observation involves two types of errors: systematic errors (which exert a non-random bias) and experimental or random errors. Systematic errors arise because of faulty control of the environment. For instance in the case of an experimental study by properly calibrating the measuring instruments and improving the experimental set-up, repetition of this type of error can be minimized.

Random errors are due to many unassignable causes and exhibit scatter as already stated. They have to be estimated. In fact the main objective of statistical analysis is to deal quantitatively with random or experimental error. Figure 6.1 shows the errr occurrence.

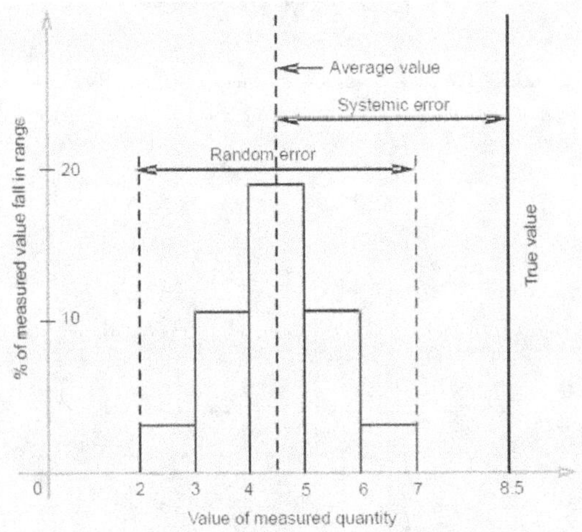

Figure 6.1 Error occurrence

The data collected from an experiment represents a sample of the population from which it was drawn. The population in this case is a collection of all possible specimens. As it is impossible to experiment with all specimens of the population, one of the main purposes of statistical analysis is to determine the best estimate of the population parameter from a randomly selected sample. While the population parameters are fixed and invariant, the parameters calculated from a sample contain random errors. Therefore, the sample provides only an estimate of the population parameter. The statistical methods, therefore, lead to conclusions having a given probability of being correct.

6.3.2 Frequency Distributions

When large number of observations is made from a random sample, a method is needed to characterize the data. The most common method is to arrange the observations into a number of equal class intervals and then determine the frequency of the observations falling within each class interval. The table below shows the yield strength of a particular material in GPa after testing 100 specimens. These set of 100 data are arranged in the form of frequency tabulation as shown below. For instance, out of 100 specimens, 5 specimens have value of yield strength in between 50 and 154 GPa.

Class interval of Y. strength	Class mid points (x_i)	Frequency (f_i)	$(x_i) \times (f_i)$	Relative frequency	Cumulative relative frequency	% of cum. relative frequency
150–154	152	5	760	0.05	5	5
155–159	157	18	2826	0.18	23	23
160–164	162	42	6804	0.42	65	65
165–169	167	27	4509	0.27	92	92
170–174	172	8	1376	0.08	100	100

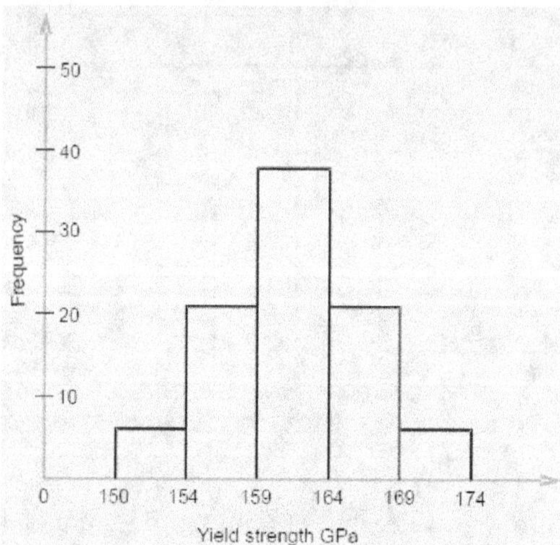

Figure 6.2 Histogram

An estimate of the frequency distribution of observations can be obtained by plotting the frequency of observations against class-interval of yields strength as a bar chart, shown in the Figure 6.2. This type of bar chart is known as "histogram". As the number of observtions increases, the size of the class interval gets reduced until we obtain a limiting curve as shown in Figure 6.3 which represents the frequency distribution of the sample. This smooth curve is a function of the random variable namely yield strength (x). This function $f(x)$ is called the probability density function. Even though the width of the shaded area in the figure is infinitesimally small dx, still the area represents the probability f x in dx. Thus,

$$P(a<x<b) = \int_a^b f(x)\,dx$$

If we know the probability density $f(x)$ for some random variable, many useful calculations can be made.

If the frequency of observations in each class interval is expressed as a percentage of total number of observations, the area under the curve which is plotted on this basis is equal to unity. If thecurv range is from $-\infty$ to $+\infty$, then,

$$P(-\infty<x<\infty) = \int_{-\infty}^{\infty} f(x)\,dx = 1.0$$

Before inferring probabilities from the frequency distribution, we need to fit the experimental results to standard statistical distributions such as normal, exponental, lognormal, etc.

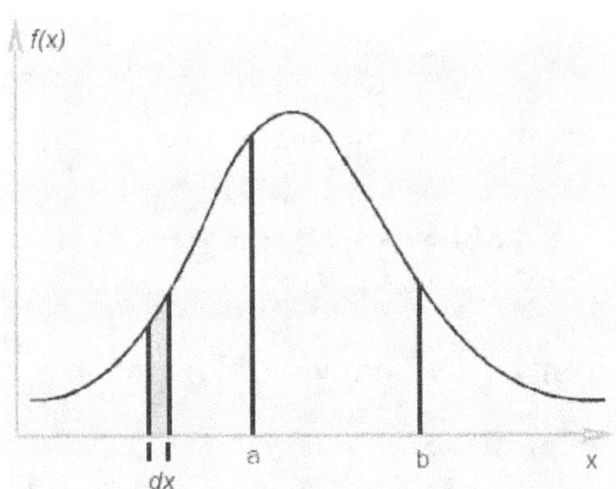

Figure 6.3 Probabiliy density functions

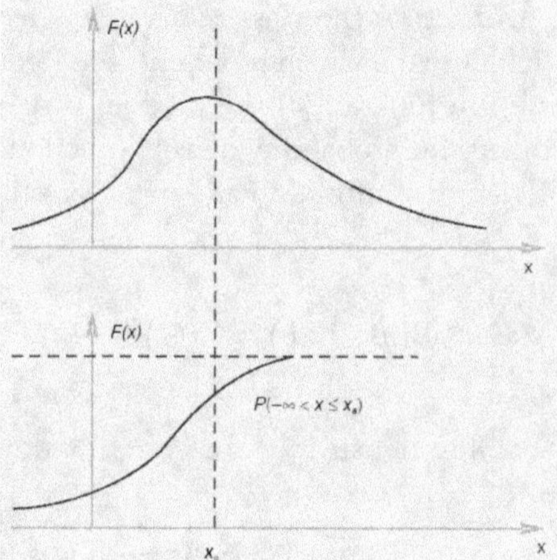

Figure 6.4 Cumulative probability density functions

Another way to present the experimental data is to arrange in a cumulative manner. The presentation of data as a cumulative distribution is preferred sometimes, because it is much less sensitive than the frequency distribution to the choice of class intervals. Thus for each probability density function $f(x)$, there is an associated function $F(X)$, called the cumulative distributio function denoted by

$$F(x) = \int_{-\infty}^{x} f(x)\,dx$$

The cumulative probability density function is shown in Figure 6.4. This indicates the probability that x is less thansome chosen value x_a.

$$P(-\infty < x \leq x_a) = \int_{-\infty}^{x_a} f(x)\,dx$$

6.3.3 Measures of Central Tendency and Dispersion

Before applying any more sophisticated statistical techniques, one must always compute two simple parameters, viz., mean and variance.

Mean   The arithmetic mean (AM) or average is the most common and important measure of the central value of any array of data. It is the best indicator of central tendency. If the observations are denoted by $x_1, x_2 \ldots x_n$, the the mean $\bar{x}$ is given by,

$$\bar{x} = \frac{\sum_{i=1}^{n} x_i}{n}$$

where, n is the number of observations.

Variance   The most important measure of dispersion of sample data is given by variance (s^2). I is calculated as

$$s^2 = \frac{\sum_{i=1}^{n}\left(x_i - \bar{x}\right)^2}{n-1}$$

where $\left(x_i - \bar{x}\right)$ is the deviation of each observation from the mean $\bar{x}$ of n observations. It is a measure of the scatter. The quantity ($n - 1$) is called the number of degrees of freedom and is equal to the number of observations minus the number of linear relations between the observations. Since the mean $\bar{x}$ represents one such relation used, the number of degrees of freedom for the variance about the mean is ($n - 1$). The other formulae for calculating variance are

$$s^2 = \frac{ü\sum_{i=1}^{n} {}_i^2 - \left(\sum_{i=1}^{n} {}_i\right)^2}{n(n-1)} \quad \text{(For ungrouped data)}$$

$$s^2 = \frac{\sum_{j=1}^{r} x_j^2 f_j - \dfrac{\left(\sum_{j=1}^{r} f_j x_j\right)^2}{n}}{n(n-1)} \quad \text{(For grouped data)}$$

where,

x_j = midpoint of group;

f_j = frequency of group;

n = total no. of observations

r = number of groups

For our example of analysing yield strength of 100 specimens, theaverage or mean value is,

$$\bar{x} = \frac{\sum_{i=1}^{n} x_i f_i}{\sum_{i=1}^{n} f_i} = \frac{16275}{ü} = 162.75\,\text{GP}_\text{a}$$

$$s^2 = \frac{\sum_{j=1}^{5} x_j^2 f_j - \dfrac{\left(\sum_{j=1}^{5} f_j x_j\right)^2}{ü}}{100-1} = 48.67$$

It is usual practice to work with standard deviation (s) which is defined as the positive square root of varance (absolute dispersion).

$$s = \left[\frac{\sum_{i=1}^{n} \left(x_i - \bar{x} \right)^2}{n-1} \right]^{\frac{1}{2}}$$

Sometimes, it is desirable to describe the variability relative to the average. If the absolute dispersion is the standard deviation (s) and the mean is ($\bar{x}$), the relative dispersion is called the "coefficient of variation" of dispersion C_v and is given by $C_v = \left[\frac{s}{x} \right]$. Suppose there are two sets of data each having s = 100, but one has $\bar{x}$ = 200 and the other $\bar{x}$ = 2000, the values for C_{v1} = 100/200 = .5 and C_{v2} = 100/2000 = .05 obviously, the second data set clustered much more closely around the mean than the first one.

Range   Range is another measure of dispersion. It is simply the difference between the largest and the smallest observation $R = (X_u - X_l)$.

There are two other common measures of central tendency other than average, viz., mode and median.

Mode   Mode (M_o) is that value of observation which occurs most frequently.

Median   The median (M_d) is the middle value of a group of observations. For a frequency distribution, the median is the value that divides the area under the curve into two equal parts.

There are certain other measures depending upon the shape of the distribution which may be useful for analysis.

Skewness   Skewness indicates lack of symmetry. A distribution is said to be skewed if mean (M), median (M_d) and mode (M_o) fell at different points, i.e., (M) $\neq$ (M_o) $\neq$ (M_d)

Karl Pearson's coefficient (S_k) is one measure used to for skewness where S_k = 3 (M – M_o) / σ where, σ = standard deviation of the distribution.

Kurtosis   Kurtosis enables us to have an idea about the flatness or peakedness of the curve. It is also known as "convexity of cuve". It is measured by coefficient β_2 or its derivative δ_2. The shapes are defined as mesokurtic (neither flat nor peaked), platykurtic (flatter than normal) and leptokurtic (more peaked). Figue 6.5 shows skewness and kurtosis.

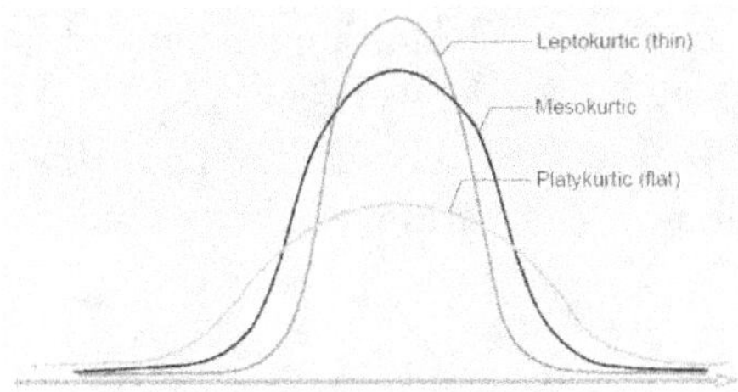

Figure 6.5 Skewness and kurtosis

6.4 POPULAR PROBABILITY DISTRIBUTIONS

We have defined the probability density functions in terms of a histogram based on real-world data. In order to design a meaningful experiment or proper analysis of a set of data, it is necessary to make a realistic choice regarding some standard distributions to be used to fit the data gathered. The reasons for using them include the generalization they provide plus the added speed and convenience of routine calculations. There are several standard distributions. Basically they are grouped under i) continuous probability distributions and ii) discrete probability distributions and iii) sampling distributions.

It is not possible to discuss all distributions and their scope of applications. To give an exposure, under the category of continuous distributions, i) Normal or Gaussian ii) Weibull and iii) Exponential will be discussed. Under the discrete distributions, i) Binomial, ii) Poisson will be covered. In sampling, *t*-distribution, chi-square and f-distribution will be covered.

6.4.1 Gaussian or Normal Distribution

Gaussian or normal distribution is the most important and commonly used distribution to fit real-world data. In addition to its goodness of fit for practical data, there are some theoretical reasons for its importance. Repeated measurements of lengths, diameter, etc., distribution of yield strength, tensile strength, etc. have been found to follow this curve. The curve is bell-shaped and symmetrical as shown in Figure 6.6. By definition, the probability density funcion for he Gaussian distribution is,

$$f(x) = \frac{1}{\sigma\sqrt{2\pi}} \exp\left[-\frac{1}{2}\left(\frac{x-\mu}{\sigma}\right)^2 \right]$$

where

$-\infty < x < \infty$ = range of x

σ = standard deviation

μ = population average value (mean)

We can choose any positive value for σ and real (+, , 0) value for μ, giving sufficient adjustability to fit many sets of practical data. For any allowed values of σ and μ, the curve is symmetrical about μ and has a total area of precisely 1.0. The value of σ decides the spread of the curve while μ locates the centre. The cumulative distribution $F(x)$ must be found by numerical integration sinc it cannot be integrated analytically.

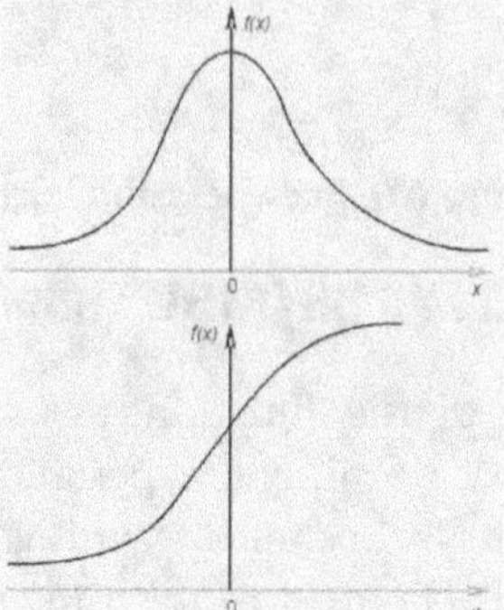

Figure 6.6 Gaussian distribution

In order to place all normal distributions on a common basis, the Gaussian distribution is frequently expressed in terms o the standard normal variable z given by

$$z = \frac{x - \mu}{\sigma} \text{ giving } f(x) = \frac{1}{\sqrt{2\pi}} \exp\left(\frac{-z^2}{2}\right)$$

For this standard normal curve, $\mu = 0$ and $\sigma = 1$. Again the total area under the standardied curve is 1.0. Cumulative distributions $F(x) = \int_{-\infty}^{z} \frac{1}{\sqrt{2\pi}} \exp\left(-\frac{1}{2}z^2\right)$ values are available in the form of table for various values of z falling between $-\infty$ and a specified value of z.

To use normal table for any calculations, we need to have numerical values μ and σ, the true mean and standard deviation of the "total population" (infinite size sample). In practice, these are never available, and we must instead use their estimates $\bar{x}$ and s, calculated from the available sample. These estimates are more reliable for larger samples and we can quantify the reliability using confidence intervals. For any variable that follows a Gaussian distribution, it may be noticed that,

68.3% of valus fall within $\pm 1\ \sigma$ of μ

95.4% of vales fall within $\pm 2\ \sigma$ of μ

99.7% of values fall within $\pm 3\ \sigma$ of μ

The use of normal distribution is illustrated by means of an example.

Example It is established that the average response of a particular phenomenon is 1.250 with standard deviation as 0.002. One experimenter conducted several repeated experiments and found that the average value varies between 1.245 and 1.255. If the values are normally distributed find the % of values that fall outside the range. We may assume

Population mean (μ) = 1.250

Sandard deviation of population (σ) = 0.002

$$z = \frac{x - \mu}{\sigma}$$

Using standard Gaussia variable, z,

$$z_u = \frac{1.255 - 1.250}{0.002} = 2.5$$

$$z_L = \frac{1.245 - 1.250}{0.002} = -2.5$$

$P(1.245 \leq x \leq 1.250) = P(-2.5 \leq z \leq 2.5)$

$= 2\ (1.0000 - 0.9938)$

$= 2 \times 0.0062 = 0.0124$

$= 1.24\ \%$

From standard normal table $P(z \leq 2.5) = 0.9938$ as the normal distribution is symmetrical, values falling outside the range will be as given.

6.4.2 Weibull Distribution

The normal or Gaussian distribution is an unbounded symmetrical distribution with long tails extending from $-\infty$ to $+\infty$. Many engineering random variables follow a bounded non-symmetrical distribution. Weibull distribution is one such distribution. It is used in many engineering problems because of its versatility. Originally used to describe the fatigue life of components, it is now widely used to describe the life of parts/components like ball bearings, gears and electronic components. One of the reasons for the popularity of this distribution is that the data can be conveniently plotted on the Weibull paper and the conformance of the data to Weibull distribution can be evaluated by the linearity of the cumulative distribution function in the same way as the normal distribution. The three parametr Weibull distribution is given in Figure 6.7.

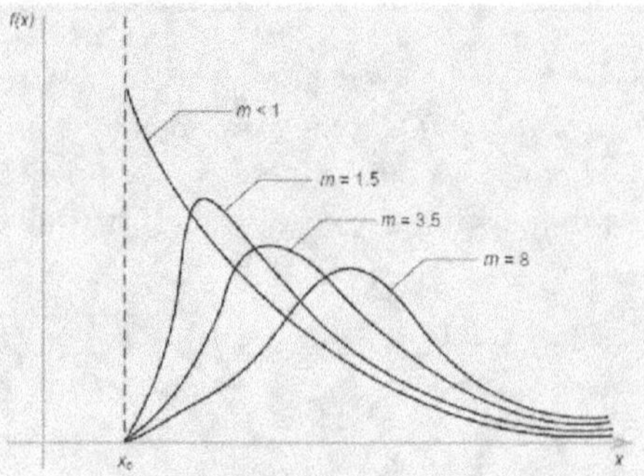

Figure 6.7 Weibull distribution

The Weibull distribution density functionis given by (for three parameter case)

$$F(x) = \frac{m}{\theta}\left[\frac{x - x_0}{\theta}\right]^{m-1} \exp\left[-\left(\frac{x - x_0}{\theta}\right)^m\right]$$

where,

θ = characteristic or scale parameter $f(x)$

m = shape or slope parameter

x_0 = expected minimum value of x and is often referred to as the threshold parameter

The cumulative distribution functin

$$F(x) = 1 - \exp\left[-\left(\frac{x - x_0}{\theta}\right)^m\right]$$

In the case of two parameters, $x_0 = 0$, then

$$F(x) = \frac{m}{\theta}\left(\frac{x}{\theta}\right)^{m-1} \exp\left[-\left(\frac{x}{\theta}\right)^m\right]; F(x) = 1 - \exp\left[-\left(\frac{x}{\theta}\right)^m\right]$$

The shape of the curve varies with change in shape parameter values, i.e., m values as shown in Figure 6.7.

6.4.3 Exponential Distribution

The exponential distribution is a special case of Weibull distribution when $m = 1$ and $x_0 = 0$. For Weibulldistribution, the probabiliy densityfunction is

$$f(x) = \frac{m}{\theta}\left[\frac{x - x_0}{\theta}\right]^{m-1} \exp\left[-\left(\frac{x - x_0}{\theta}\right)^m\right]$$

Putting $m = 1$ and $x_0 = 0$

$$f(x) = \frac{1}{\theta}\exp-\left(\frac{x}{\theta}\right) = \frac{1}{\theta}e^{-x}\Big/\theta$$

Putting $\lambda = \dfrac{1}{\theta}$ the exponential distribution becomes,

$$f(x) = \lambda e^{-\lambda x} \text{ ; for } x > 0 \text{ and } \lambda > 0$$

The shape of exponential distribution is shown n Figure 6.8.

The mean of exponential distribution and variance $(\bar{x}) = \dfrac{1}{\lambda}$ and variance $(s^2) = \left(\dfrac{1}{\lambda}\right)^2$

$$\therefore \ s = \frac{1}{\lambda}$$

In this distributionboth mean and standard deviation are equal, i.e., $\bar{x} = s$.

The cumulative disribution function of exponential distribution is,

$$F(x) = 1 - e^{-\lambda x} \text{ ; } x > 0$$

Exponential distribution plays a key role in the theory of queues and in reliability analysis. This is used to represent activities where most of the events take place in a relatively short time, while there are a few which take very long time. The service times in queuing systems, inter-arrival time of vehicles in a highway, the life of some electronic components are some of the examples where exponential distribution could be used. This distribution is beter suited to model failure of the complete system.

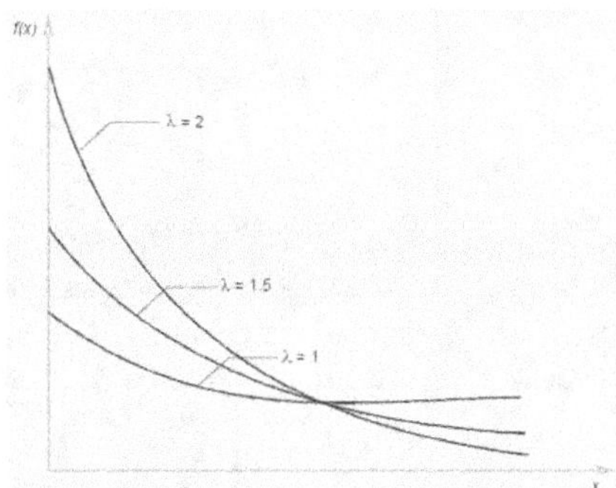

Figure 6.8 Exponential distribution

6.4.4 Binomial Distribution

In continuous distribution like Gaussian, the probability of the random variable taking on any specific value is zero, since probability is represented by an area within an interval. For those applications where such probabilities are non-zero, we use discrete distributions. Binomial and Poisson are most commonly used discrete distributions.

Binomial distribution applies to situations where events are judged on a" "yes" or "no" basis. In quality control after inspection, a part is declared as defective or not defective. We assume that the parts come from a population which has fixed percentage of good and defective items, and this percetage remains the same as we draw samples to be tested.

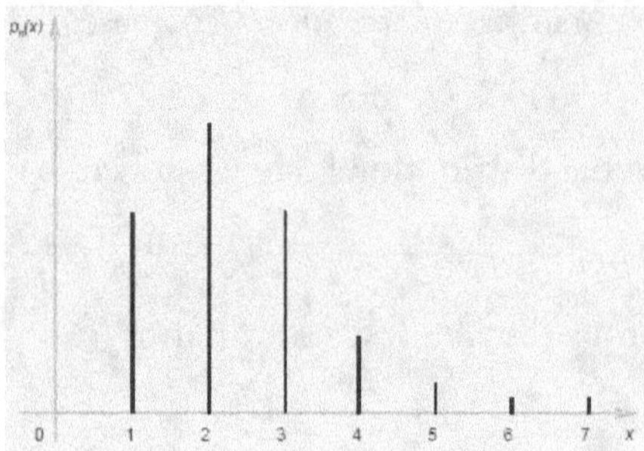

Figure 6.9 Binomial distribution (mass function)

Let the occurrence of selecting a good part is "success" and its non- occurrence as "failure". Let p denote the probability of success and q denote the probability of failure such that $p + q = 1$. The number of successes in n trails may be 0, 1, 2, 3,... $r...n$ and obviously a chance variate. In this case, n represents the number of parts in a batch being tested. The probability of x successes in n trials is given by $P_n(x) = p^x q^{n-x} \dfrac{n!}{x!(n-x!)} P_n(x)$ is

the probability of geting x good parts in a batch of n. It is also written as $p(x) = nC_x p^x q^{n-x}$. This is the probability mass functon of binomial distribution. The mean of the distribution $\bar{x} = n.p$ and variance $(\sigma^2) = n.p.q$. The probability of etting x or less good parts in a batch of n is given by

$$C_n(x) = \sum_{x=0}^{x} p_n(x)$$

In practical applications, we are often in the position of not knowing the true value of p and we try to estimate it from data on a finite size sample. Figures 6.9 and 6.10 shows Binomial distributio mass function and cumulative distribution respectively.

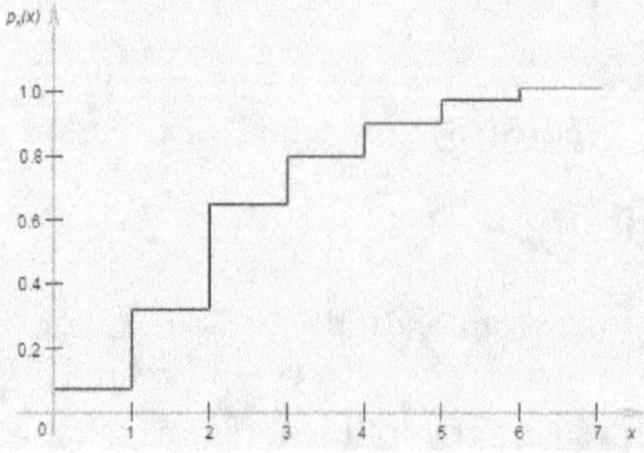

Figure 6.10 Binomial cumulative distribution

6.4.5 Poisson Distribution

Poisson distribution is the discrete version of the exponential distribution. Poisson distribution is used to model the number of independent events that occur in a fixed amount of time or space. This distribution finds application in a wide variety of situations like number of typographical errors in a page, number of defects or flaws in a square metre of cloth, number of alpha particles emitted by a radioactive substance, etc. The probability mas function of Poisson distribution is given in Figure 6.11.

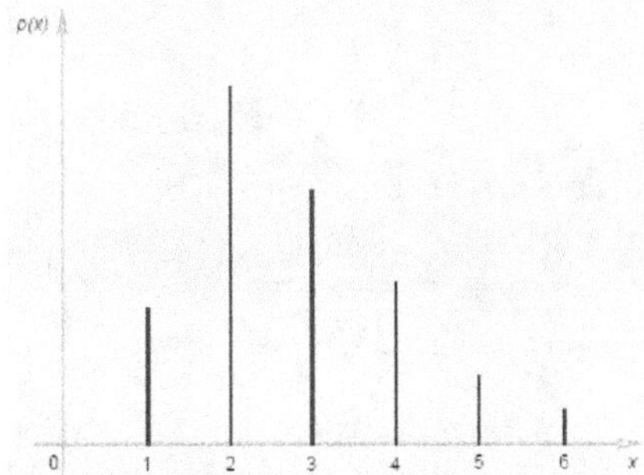

Figure 6.11 Poisson distribution probability mass function

If λ represents the average number of occurrence of an event, the probability of occurrence of x events is given by

$$p(x) = \frac{e^{-\lambda}\lambda^x}{x!}$$

for $x = 0, 1, 2, 3...n$ and $x > 0$. The mean and variance of Poisson distribution are equal, i.e., mean $\bar{x} = \lambda$ and σ^2. The cumulative probability distribution is given by $C(x) = \sum_{i=0}^{x} e^{-\lambda}.\lambda^i$.

The use of Poisson distribution is illustrated using an example.

Example   The surface defects noticed by the quality control department in each 10 m length of cold-drawn bar stock in number is given below. The prescribed quality specification is a maximum of 3 defects per 10 m length. What is the probability of being or eyond the specification limit of 3 defects per 10 m length?

Defects/10 m	Nos.	Defects/10 m	Nos.
0	40	4	1
1	9	5	1
2	4	6	1
3	2		

The average number of defects is clulated as given below.

Defects	No.	Total
0	40	(0) (40) = 0
1	9	(1) (9) = 9
2	4	(2) (4) = 8
3	2	(3) (2) = 6
4	1	(4) (1) = 4
5	1	(5) (1) = 5
6	1	(6) (1) = 6
	58	38

The average number of defects is $\lambda = \dfrac{38}{58} = 0.655$

$$e^{-\lambda} = e^{-0.655} = 0.519$$

$$P(0) = \frac{(0.519)(0.655)^0}{0!} = 0.519$$

$$P(1) = \frac{(0.519)(0.655)^1}{1!} = 0.339$$

$$P(2) = \frac{(0.519)(0.655)^2}{2!} = 0.111$$

$$P(3) = \frac{(0.519)(0.655)^3}{3!} = 0.024$$

Probability of being at specification limit of 3 defects/length $P(3) = 0.024$.

Prbability of being beyond the specification limit = $P(x > 3)$

i.e., $P(x > 3) = 1 - P(x \leq 3)$

$$= 1 - [P(0) + P(1) + P(2) + P(3)]$$

$$= 1 - [0.519 + 0.339 + 0.111 + 0.024]$$

$$= 1 - 0.993$$

$$P(x > 3) = 0.007$$

6.5 SAMPLING DISTRIBUTIONS

6.5.1 Distribution of Sample Means

The major problem in statistics is relating the population and the samples that are drawn from it. There are two main aspects in this: i) What the population tells us about the behaviour of samples drawn from it, ii) What the sample or series of samples tell us about the population from which the samples were drawn? We discuss the first question.

Let us assume we have a population of N items, we take all possibl samples size of n = 10 and determine the mean of each sample $\bar{x}$. This would give a population means of $\bar{x}$'s, one for each sample of size n. We call this as sample statistic. Similarly, we can have sampling distribution for s^2 or for any other sample statistic. The distribution of sample statistics are called "sampling distributions".

The sample mean provides an unbiased estimate of the mean of the population from which the sample was drawn and will be normally distributed about the population mean, μ. We denote the mean of the distribution of $\bar{x}$ as $\mu_{\bar{x}}$ and it is equal to the mean of the population of x's (i.e., $\mu_{\bar{x}} = \mu$) if the population is very large. (i.e., n > 30). The standard deviation of the sampling distributon is called the standard error of the mean and it isequal

to $\sigma_{\bar{x}} = \dfrac{\sigma}{\sqrt{n}}$. We can use this to establish the standard variable z, $z = \dfrac{\bar{x} - \mu}{\sigma_x} = \dfrac{\bar{x} - \mu}{\sigma_x/\sqrt{n}}$ which is normally distributed with mean '0' and standard deviation 1.

6.5.2 Student's *t*-Distribution

In many situations we may not know the standard deviation(σ) of the population. In such cases, we cannot substitute s for σ in the equation of the previous para and assume that the statistic is normally distributed unless n is greater than 30. However, if x is approximately normally distributed and $n < 30$, the sampling distribution for the meanis the student's

t-distribution. The statistic t is given by $\dfrac{\bar{x} - }{s/\sqrt{n}}$ for $v = n - 1$ degrees of freedom.

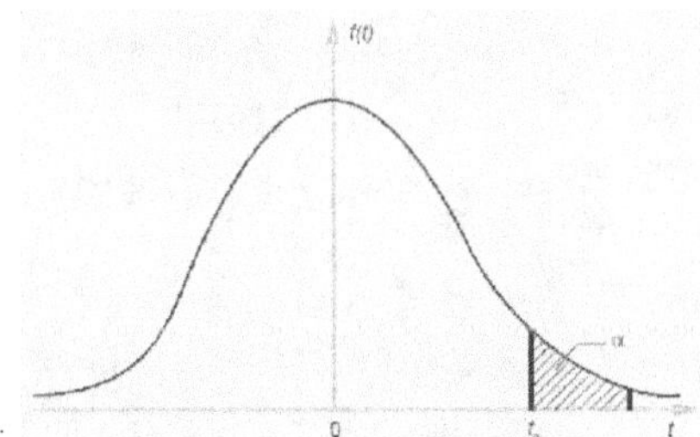

Figure 6.12 Student's t-distribution

The t-distribution is symmetrical about the line $t = 0$ and the maximum ordinate is at $t = 0$. There is a different curve for each value v. As v increases, the distribution of t approaches normal distribution. The number of d.f. of a statistic denoted by v is defined as n (independent observations) in the sample minus the number of population parameters which must be estimated from the samples. In the present case, s is the only statistic and hence $v = (n - 1)$. The t-distribution values are available in tables where for a given v and probability α, the t_0 value can be found. This gives $p(t >, t_0) = \alpha$ the level of significance. The shape of t-distribution is shown in Figure 6.12 and it is similar to a normal distribution.

6.5.3 Chi-square Distribution

The distribtion of sample variances s^2 from a norma population with variance σ^2 is given by chi–square (χ^2) distribution.

$$\chi^2 = \frac{(\nu)(s^2)}{\sigma^2}$$

$$= \frac{(n-1)s^2}{\sigma^2}$$

The chi–square statistic, χ^2 can also be computed for a given set of data by computing the observed frequency of data in each class interval and the expected frequency for the sme class interval, predicted by the theoretical distribution as to

$$\sum_{i=1}^{k} \frac{(O_i - E_i)^2}{E_i}$$

where,

O_i = observed frequency,

E_i = expected frequency,

k = number of class intervals.

If ($\chi^2 = 0$) the the observed and theoretical frequences agree exactly, whereas if ($\chi^2 > 0$) they do not. The larger the value of χ^2 the greater is the discrepancy. The chi–square values are tabulated and availabe as published data. The χ^2-statistic is tabulated by d. f. vs. $(1 - \alpha)$ or the significance level. This is extensively used to establish goodness of fit test between theoretical distributions such as normal, binomial, etc. and the empirica l distibutions obtained from sample data.

6.5.4 *F*-Distribution

Let N_1 (μ_1, σ_1) and N_2 (μ_2, σ_2) be two normally distributed populations with mean μ_1 and μ_2 and variances σ_1 and σ_2 respectively. If we take two independent samples of sizes n_1 and n_2 from the two populations respectively with variances of samples as s_1^2 and s_2^2, the F-distribution tells us whether the two samples come from populations having equal variances. The larger of the two sample variances must be the numeraor and the smaller, the denominator. For instance s_1 is larger, then

$$F_{\nu_1 \nu_2} = \frac{s_1^2}{s_2^2}$$

where, $\nu_1 = (n_1 - 1)$ and $\nu_2 = (n_2 - 1)$ are the degrees of freedom. If we make simple calculation of s_1 and s_2 and make comparison of s_1 and s_2, it can be misleading. The values of F-distribution is available in the form of table for different values of v_1 and v_2 and chosen value of α, the level of significance. From the table value the critical value of F is known and by comparing it with the calculated F-value the decision is made.

6.6 FITTING OF THEORETICAL DISTRIBUTIONS

6.6.1 Graphical Method

Fitting a theoretical distribution like normal or Poisson to the observed response variable or dependent variable has several advantages. Not only this has made parameter estimation for the theoretical distribution easy, but also has enabled the study of behaviour pattern. Further, generation of simulated data from the theoretical distribution fitted is easy. Many phenomena in real-life situations relating to engineering follow normal or Gaussian distribution. Therefore using the observed data, how one can check graphically whether it fits to a normal distribution or not is explained below.

First a rough comparison of the cumulative distribution (CDF) of the observed data is made with CDF of the normal distribution. Based on rough guess that the observed data approximates to a normal distribution, a cumulative plot on a normal probability paper is made. First the mean ($\bar{x}$) and standard deviation (s) of the observed data is calculated. Then plot a point P_1 at ($\bar{x}$, 0.5) and another point P_2 at ($\bar{x}$ + s, 0.84) on the normal probability paper. Through P_1 and P_2, draw a straight line. Thereafter, the individual observed points are plotted on the paper as shown in Figure 6.13. The degree of fit to the normal distribution is determmned by noting how well the data clusters around the straight line drawn.

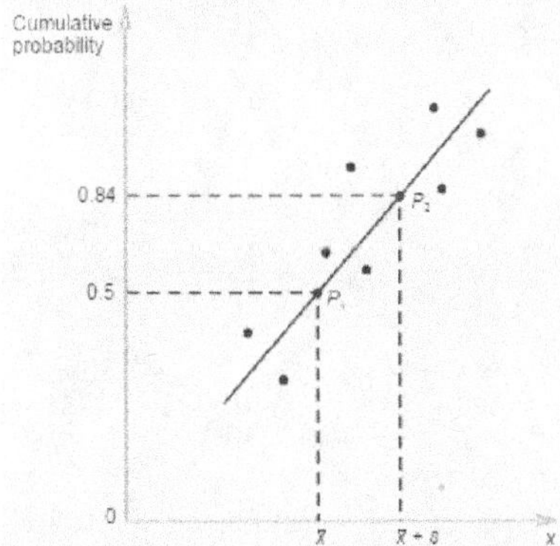

Figure 6.13 Normal probability plot of data

6.6.2 Goodness of Fit Test

The goodness of fit test is a statistical approach to check whether the observed or empirical data fits to an assumed theoretical distribution. In other words, it enables to draw conclusion that the observed data belongs to the assumed distribution. For this purpose the commonly used statistical fitness tests are (i) chi-square test, and (ii) Kolomogrov–Smirnov test. Here, the use of chi-square test is explained by an example.

Example   A researcher made 100 observations of a phenomenon from his exerimental set-up and prepared a frequency table of the data as given below:

Range	Frequency	Range	Frequency
60–62	5	68–71	27
63–65	18	72–74	8
66–68	42		

Based on his knowledge about the phenomenon, he perceives that it should follow a normal distribution. Show how he will statstically establish his perception.

The researcher first calculated the mean $\bar{x}$, and the standard deviation s for his data set.

hen the z value for each class boundary was calculated based on the relation $\frac{x - \bar{x}}{s}$. Using the normal table the area under normal curve from 0 to z for the class boundary and then the area for each class interval was arrived at.

From the area for each class interval, the expected frequency wa arrived at as shown in the table below:

Range	Class boundary	z for class boundary	Area under normal curve	Area for class	Expected frequency	Observed frequency
	59.5	–2.72	0.4967			
60–62				0.0413	4.13	5
	62.5	–1.70	0.4554			
63–65				0.2068	20.68	18
	65.5	–0.67	0.2486			
66–68				0.3892	38.92	42
	68.5	0.36	0.1406			
69–71				0.2771	27.71	27
	71.5	1.39	0.4177			
72–74				0.0743	7.43	8
	74.5	2.41	0.4920			

From the above table the chi–square (χ^2) can be calculated using the relation $\chi^2 = \sum_{i=1}^{R} \frac{(O_i - E_i)}{E_i}$ and the degrees of freedom, ν is calculated based on the number of cells and nmber f parameters (m) used. In this case the number of cells is 5 and $m = 2$.

$\therefore d.f.(v) = k - 1 - m = 5 - 1 - 2 = 2$

and

$$\chi^2 = \frac{(5-4.13)^2}{4.13} + \frac{(18-20.68)^2}{20.68} + \frac{(48-38.92)^2}{38.92} + \frac{(27-27.71)^2}{27.71} + \frac{(8-7.43)^2}{7.43}$$

$$= 0.1832 + 0.3473 + 0.2437 \ \ 0.0187 + 0.0437$$

$$\chi^2_{cal} = 0.8360$$

From the table of chi–square values or $v = 2$ and 95% confidence level, the critical value noted is 5.99. For $v = 2$, $\chi^2_{0.951} = 5.99$. Since the calculated χ^2 is less than the critical value of 5.99, the researcher can very well conclude that the fit of observed data with normal distribution is very good.

7
Sample Design and Sampling

7.1 SAMPLING DESIGN

7.1.1 Sampling

All items of interest such as individuals or of their attributes or results of operations, which are subject to investigation, are termed as "population" or "universe" in statistics. Thus in statistics population represents an aggregation of objects, animate or inanimate, under study. The population may be finite or infinite. A part or a small portion selected from the population is called sample and the process of such selection is called "sampling". Sampling is resorted to when it is not possible to enumerate the whole population or when it is too costly to enumerate in terms of money and time. If the analysis is to be purposeful, then sampling should be unbiased and representative.

One should not be confused with terminology such as parameter and statistics. Normally we use the term parameter when we talk of population. The population quantities such as "population mean" and "population variance" are called "population parameters" or simply "parameters" and are represented by letters μ and σ^2 respectively. Statistical measures computed from sample observations (i.e., outcome of experiments, since the outcome of each experiment is a sample point) such as sample mean and sample variance are represented by $\bar{x}$ and s^2 respectively, are termed as "statistics". Generally if a researcher tries to generalize the results based on repeated trials in different contexts, then he/she really refers to the population parameter after generalization.

The aim of the sampling exercise is to get as much information as possible—ideally the whole information about the population from which the sample is drawn. In particular, given the form of parent population, we would like to estimate the parameters of the population or specify the limits within which the population parameters are expected to lie with a specified degree of confidence from study of samples. We have to clearly understand that the logic of the theory of sampling is the logic of induction. That is, we move from particular (sample) to general (population) and hence results will have to be expressed in terms of probability.

7.1.2 Sample Design

A sample design is a definite plan for obtaining a sample from a given population. It refers to the technique or the procedure the researcher would adopt in selecting items for the sample and the size of the sample. Sample design is determined before data are collected. There are many sample designs from which a researcher can choose. Some are relatively more precise and easy to use than others. A researcher must select/prepare a sample design which should be reliable and appropriate for his/her research study.

While developing a sampling design, the researcher must pay attention to the following aspects.

- The type of universe, whether finite or infinite and also clearly define the set of objects that constitute the universe.

- Sampling unit should be decided before selecting the sample. For example the unit may be state, district or a zone, etc. or specific items or individuals.

- Source list or sampling frame from which sample is to be drawn has to be prepared.

- Determine the size of sample or the number of items to be selected from the universe keeping in mind the desired precision and the chosen confidence level for the estimate.

- In deciding sample design one must consider the question of specific population parameters which are of interest such as mean or other measures.

- Cost considerations from practical point of view have a major impact relating to not only the size but also the type. Hence budgetary constraint is an important issue.

- The researcher should decide the type of sample and the technique to be used in selecting the items, which gives smaller sampling error for given sample size and cost.

Sampling errors represent the random variations in the sample estimates around the true population parameters. They occur randomly and are equally likely to be on either direction. Their nature is generally of compensatory type and the expected value of such errors is equal to zero. Sampling error is the sum total of frame error, chance error and response error. It decreases with increase in sample size and it happens to be of a smaller magnitude in the case of a homogeneous population. Sampling error can be measured for a given sample design and size. The measurement of sampling error is usually called the "precision of the sampling plan". While selecting a sampling procedure, the researcher must ensure that the procedure causes relatively small sampling error and also helps to control the systematic bias in a better way.

7.2 TYPES OF SAMPLE DESIGNS

There are different types of sample designs and they are based on two factors, viz., the representation basis and the element selection technique. On the representation basis, they are classified as probability sampling and non-probability sampling. Probability sampling is based on the concept of random selection whereas non-probability sampling is non-random sampling. On element selection basis, the sample may be either unrestricted or restricted sampling. When a sample element is drawn from a large population, then the sample so drawn is known as "unrestricted sample" whereas all other forms of sampling are covered under the term "restricted sampling". Thus, the sample designs are basically categorized into two types—non-probability and probaility sampling.

Representation basis / Element basis	Probability sampling (1)	Non-probability sampling (2)
1. Unrestricted sampling	Simple random sampling	Haphazard sampling or convenience sampling
2. Restricted sampling	Complex random sampling such as cluster, stratified, systematic, etc.	Purposive sampling such as quota, judgement sampling.

7.2.1 Non-probability Sampling

Non-probability sampling does not afford any basis for estimating the probability that each item in the population has a chance of being included in the sample. This sampling is also known as purposive, deliberate and judgement sampling. In this, the sample items are selected deliberately by the researcher using his/her choice. In such a design, personal bias has a great chance of entering into selection of the sample. The investigators may select a sample in such a way that it yields results favourable to their point of view and if that happens, the entire inquiry may get vitiated. Thus, there is a danger of bias entering into this type of sampling technique.

The experience of the investigators, their impartiality and capability of taking sound judgement may result in selecting a sample which is tolerably reliable. Sampling error in this type of sampling cannot be estimated and the element of bias, great or small, is always there. This type of sampling technique is usually adopted in small inquiries and researches by individuals due to the relative advantage of time and money inherent in this type. Quota sampling is an example of non-probability sampling. This is used in surveys. Here, the interviewers are simply given quota, and selection of a person for interview is left to their discretion. Quota samples are essentially judgement samples and inferences drawn on their basis are not amenable to statistical treatment in a formal way.

7.2.2 Probability Sampling

Probability sampling is also known as random sampling or chance sampling. Under this, every item of the population has equal chance of being included in the sample. The results obtained from random sampling can be assured in terms of probability. That is, one can measure the errors of estimation or significance of results obtained from a random sample. This method is much superior over the deliberate sampling design. In brief, the implications of a random sampling (or simple random sample) are:

- Each element in the population has an equal probability of being selected into the sample and all choices are independent of one another.
- Each possible sample combination has an equal probability of being chosen.

Suppose a population has a finite size of 5 numbers and we want to select a sample size of 2 from it, then there are $5C_3 = 30$ possible distinct samples of the required size. If we choose any one of the samples in such a way that each has the probability of 1/30 of being chosen, we will then call this a random sample.

The method of obtaining a random sample can be simplified in actual practice by the use of random numbers table. Suppose we are interested in taking a sample size of 10 units from a population of 3000 units. We can number each item from 2001 to 5000. We can then select from the random number table random numbers which are not less than 2001 and not greater than 5000.

A portion of the random nmber is given below.

1772	4640	2331	9044	6621	2898
4033	3491	3587	6568	1960	1387
5939	0173	2078	5424	1645	5870
0586	2133	5797	5406	1041	6707
3585	9353	1938	2322	6759	5659

If we decide to read the table numbers randomly from left to right, starting from the first row itself, we obtain the following numbers: 4640, 2331, 2898, 4033, 3491, 3587, 2078, 2133, 3585, and 2322 which are >2001 and < 5000. The units bearing the above serial numbers form the required random sample.

7.2.3 Complex Random Sample Design

Probability sampling under restricted sampling techniques constitute complex random sample designs. They may also be called "mixed sampling designs" as many such designs represent a combination of probability and non-probability procedures in selecting a sample. Some of the popular sampling designs are briefly discussed below.

Systematic sampling   If in a sampling, we select every i[th] item on a list, then this constitutes systematic sampling. An element of randomness is introduced in this by using random numbers to pick up the starting unit. For instance, if a 5% sample is desired, the first item would be selected randomly from the first 20 units, and thereafter every 20th unit would be automatically included in the sample. Though this is not a random sample in the strict sense, it is often treated as a random sample.

Stratified sampling   If a population from which a sample is to be drawn does not constitute a homogeneous group, stratified sampling technique is generally applied in order to obtain a representative sample. In this method, the population is divided into several sub-populations that are individually more homogeneous than that of the total population, and then we select items from each stratum to constitute a sample. If each stratum is of different size, proportionately the number of items to be selected from each strata is made. Within each stratum simple random sampling approach is adopted.

Cluster sampling   In cluster sampling the total population is divided into a number of relatively small sub-divisions which are themselves clusters of still smaller units. Then some of these clusters are randomly selected for inclusion in the overall sample. Suppose we want to estimate the proportion of defective parts in a population of 25,000 units packed in 500 cases each combining 50 parts, we would consider the 500 cases as clusters and randomly select n cases from the 500 and examine all the parts in each randomly selected case. If clusters happen to be some geographic sub-divisions, then the cluster sampling is known as "Area sampling".

Multistage sampling   This sampling is a further development of cluster sampling and is applied in big investigations like a household survey extending to a large geographical area, say the entire city. The first stage is to select a large primary sampling unit such as zone in the city. Then we select certain division and do the survey. This represents a two-stage sampling design, with the ultimate sampling units being the cluster of divisions. Only important aspects with regard to different types of sampling is covered here.

7.3 THE STANDARD ERROR

The standard error (SE) of a sampling distribution is nothing but its standard deviation. This is an important concept in sampling analysis. Using the standard error one can test whether the observed distribution based on experimental data differs significantly from the expected frequency distribution. Normally, if the difference is less than thrice the standard error (3SE), the difference could be considered as due to chance causes. If it is greater than 3SE, the difference is due to some assignable causes. The above decision is based on the concept that $\bar{x} \perp 3SE$ accounts for 99.73% of the area under the normal distribution, which is a bell-shaped symmetrical curve

The standard error is an important measure in significance test in deciding the hypothesis using the sample data. At 5% level of significance ($\infty = 0.05$) if the estimated parameter differs from the calculated statistic by more than 1.96SE, the difference is considered as significant at 5% level. The figure, 1.96, is the critical value assuming normality. This implies that the difference lies outside the limits (i.e., it lies in the 5% area) and we can say with 95% confidence, that the said difference is not due to fluctuation in sampling. The "sampling error" at a particular level of significance is estimated by the product of the critical value and the standard error (SE).

By using the standard error, we can estimate the reliability and precision of a sample. If the sample distribution is uniform, the SE will be smaller and reliability will be high. The size of the standard error depends upon the sample size to a greater extent and varies inversely with the size of the sample. Further, the standard error enables us to specify the limits within which the parameters of the population are expected to lie with specified degree of confidence. The table below gives the percentage of samples having their mean values within the range of population mean $\mu \pm SE$.

Range	Percentage values
$\mu \pm 1\ SE$	68.27
$\mu \pm 2\ SE$	95.45
$\mu \pm 3\ SE$	99.73
$\mu \pm 1.96\ SE$	95.00
$\mu \pm 2.57\ SE$	99.00

7.4 SAMPLE SIZE FOR EXPERIMENTS

7.4.1 Confidence Interval Approach

Selection of proper sample size is an important aspect while conducting experimental study. The evaluators of a thesis raise the question of statistical validity of repetition of experiments and the confidence attached to the results of the experiments. Therefore, it is imperative for the researcher to estimate the sample size required for an assumed level of confidence. In case the size of sample has been assumed, the analysts have to make estimation of the confidence limits that they could attach for the given level of confidence.

Confidence interval The result obtained from an experiment gives a sample value and when the experiment is repeated, a number of sample points are obtained. Based on the sample values the sample average $\bar{x}$ and sample standard deviation(s) are calculated. When sample values are used to estimate universe or population parameter, a measure of how much confidence can be placed in them is needed. This is answered through the use of confidence limits or confidence interval based on the sample values.

Confidence limits are placed so that we are $(1 - \alpha)$ per cent sure that an interval based on the sample values includes the population value being estimated. The choice can be at will. This interval estimation or confidence limits estimation is commonly used to make probability statements about population from which a sample has been drawn or to predict results of future sample from the same population.

The analysts determine the confidence limit for a parameter related to their study or experiment so that they can attach a specified degree of confidence that the parameter lies within the interval. For example if they determine 95% confidence limits ($\infty = 0.05$) of the mean of a particular parameter, then in the long run when experiments are repeated, the mean of the population will lie within the limits specified for at least 95% of the time. The confidence intervals can be established in several ways statistically.

If the analysts are aware of the variance of the parametr, in other words the population variance (σ^2), then the confidence interval on the mean of the parametr can be etimated using the relation,

$$\left(\bar{x} - z_{\alpha/2}\,\sigma/\sqrt{n}\right) \leq \mu \leq \left(\bar{x} + z_{\alpha/2}\,\sigma/\sqrt{n}\right)$$

where,

($\bar{x}$) is the mean of sample,

μ is the population mean.

$z_{\alpha/2}$ is the critical value obtained ∝rom normal table for the assumed value of ∞ and

n is the sample size.

In many circumstances, the analysts may not be aware of the variance of the population. In that case they can use the sample variance (s^2) and use t-distribution with $n - 1$ degrees of freedom for calculting the confidence limits as given below.

$$\left(\bar{x} - t_{\alpha/2}\,s/\sqrt{n}\right) \leq \mu \leq \left(\bar{x} - t_{\alpha/2}\,s/\sqrt{n}\right)$$

Similarly confidence interval for sample variances can also be attached when experiments are repeated. The distribution of sample variances s^2 from a normal population with variance σ^2 is given by the chi–square distribution. Using this, the confidece interval on variance can be computed as,

$$\left((n-1)s^2/\chi^2_{1-\alpha/2}\right) \leq \sigma^2 \leq \left((n-1)s^2/\chi^2_{\alpha/2}\right)$$

For an assumed value of α, the can be taken from the chi-square table.

7.4.2 **Repetition of Experiments**

An important question in sampling is what should be the sample size (n)? Assuming each repetition of the experiment under the same conditions gives one value; the question is how many times the experiment has to be repeated. To answer this, we require knowledge of i) the nature of population, ii) the type of sampling, iii) the nature of study, iv) the standard of accuracy or precision desired, v) the cost of conducting each experiment, vi) the cost of imprecise results, vii) the variability of the process, and viii) the amount of confidence required for the results of the experiment. Neglecting cost factors, the confidence interval we discussed earlier allow us to place the limits and determine the sample size required or the number of times an experiment is to be repeated to get the desired confidence (1–$\propto$) for these limits. As a general rule the sample must be of an optimum size.

The sample size n for the confidence level (1–$\propto$), with the interval on either side of the mean as $\pm d$, is calculated using the relation given below. The d denotes the error range that decreases with increase in sample size. Therefore, for a given error size one can estimate the sample size required for the experiment. The experimenter will have make a reasonable assumption of the value for d.

$$n = \left[\frac{(z_{\alpha/2})(\sigma)}{d} \right]^2 \quad \text{if the variance of the population is known.}$$

Suppose we want to determine sample size with 95% confidence and with an assumed value for d as ± 0.1 of the population mean. Knowing that the value of σ is 0.2, the value of n can be calculated as

$$n = \left[\frac{(1.96)(0.2)}{(0.1)} \right]^2 = 15.4 \text{ or } 16$$

To get the precision required, the sample size is 16. The $z_{\alpha/2}$ is the standardized normal statistic for the probability we seek.

In many cases, the exact value of σ may not be known. Suppose we have some idea of the highest and lowest possible results of the experiment, (feasible range), we can assume the range is approximately equal to 4σ to estimate σ. If the variance is not known, pilot experiments may be conducted and based on the results, an estimate of variance (s^2) can be computed. We can use t-distribution and the sample size can be estimated assuming a reasonable value for d.

$$n = \left[\frac{(t_{\alpha/2})(s)}{d} \right]^2$$

where, t is the tabulated t–value for the desired confidence level and degrees of freedom of initial sample and d is half–width of desired confidence interval assumed.

7.5 RUN SIZE FOR SIMULATION EXPERIMENTS

7.5.1 Need to Determine Run Length

When we use simulation to study a stochastic system, we represent one or more of the variables in the model by probability distributions from which we draw samples. Since these samples are randomly drawn, we have some degree of imprecision in the result, which is highly influenced by the choice of the sample size. Since we use the information furnished by the simulation experiment as the basis for decision regarding the operation of the real system, we want this information to be as accurate and precise as possible or at least we want to know the degree of imprecision present. It is therefore essential that a statistical analysis be conducted to determine the required sample size.

The sample size or run length may be determined in either of the two ways: i) prior to and independently of the operation of the model; ii) during the operation of the model and based upon the results generated by the model.

7.5.2 Prior Determination Approach

It is frequently possible to carry out certain forms of prior analysis based upon the knowledge of the model. Many forms of analysis are based upon the assumption that the responses of the model are independent and normally distributed. These assumptions are justified because of the central limit theorem from probability theory. It is usually sufficient that the response is the additive sum of a large number of contributing effects. The significance of central limit theorem lies in the fact that it permits us to use sample statistics to make inferences about population parameters without knowing the nature of the population, other than what we get from the sample.

In the most straightforward case, where we can invoke the central limit theorem and assume no auto-correlation, we can take a confidence limit approach to determine the sample size required for estimating the parameters to a specified level of precision. These parameters are population mean, standard deviation, etc.

Suppose we with to determine an estimate of $\bar{x}$ of the true population μ, such that, $P[\mu\text{-d}) \leq \bar{x} \ (\mu\text{-d})] = (1-\alpha)$ where is the sample mean, is the ppulation mean and is the probability that the interval $\mu \pm$ d contains $\bar{x}$. The problem is to determine the sample size such that the above equation holds. If our assumptions of normality are valid, we can show that, $n = \left[\dfrac{(z_{\alpha/2})(\sigma)}{d} \right]^2$ where $z_{\alpha/2}$ is the two-tailed standardized normal statistic for the probability we seek. In many cases, we must either guess at the value of σ or run a short pilot run. As in the case of experiments, if we have some idea of what the highest and lowest possible responses of the system might be (feasible range), we can assume that the range is approximately equal to 4σ to estimate σ.

If the variance is unknown, then we can conduct a pilot run or take a preliminary sample and obtain an estimate of the variance (s^2), then compute the total number of observations necessary. Thus, when σ is unknown, the run length can be estimated using t–distribution, $n = \dfrac{t^2 s^2}{d^2}$ where, t = tabulated t-value for the desired confidence level and d. f of initial sample, d = half–width of desired confidence interval and s^2 is the estimated variance from sample.

The example given below illustrates how the run length can be determined using the concepts discussed above.

Example   An analyst who simulates a chemical process wishes to have a 95% probability that the error of estimate of the output parameter of interest is less than ± 4 tons. The information gathered indicates that the range of output covers 72 tons. i) Determine how he will arrive at the run length for his simulation. ii) In case he is not aware of the possible output range how will he decide the run length?

i. The estimated feasible ange f output is 72 tons. Therefore, he can take it that 4σ = 72 and $\sigma = 72/4 = 18$. Since he wants the limits to be ±4 ton, $d = 4$. From the normal table for 95% proability, i.e., $\alpha = 0.05$, $z_{\alpha/2} = 1.96$. Therefore, the run size n is

$$n = \left[\frac{(z_{\alpha/2})(\sigma)}{d} \right]^2$$

$$= \left[\frac{(1.96(18))}{4} \right]^2 = 78$$

ii. Since he is not ware of the range of output, he may not have any idea of . Instead of setting an arbitrary value for d, he an assume that his estimate should lie within the interval $\mu \pm \dfrac{\sigma}{4}$ with probability of 0.95.

Thus, he can take $d = \dfrac{\sigma}{4}$

$$n = \left[\frac{(1.96)(\sigma)}{\left(\frac{\sigma}{4}\right)} \right]$$

$$= \left[\frac{(4)(1.96)(\sigma)}{(\sigma)} \right]^2$$

$$= \left[(4)(1.96) \right] = 61$$

7.5.3 Use of Automatic Stopping Rule

In this approach the confidence intervals for output values are determined during a simulation run and then the execution is terminated when a predetermined confidence level is reached. Thus, we can avoid the inefficiencies of runs that are either too long or too short. One of the approaches to incorporate automatic stopping rule is to run the simulation in two stages. First run assumed sample size n, use the results to estimate n^* (desired sample size) by one of the methods previously described. If $n^* < n$, the run is over, otherwise, extend the run by $(n^* - n)$.

8
Hypotheses and Anova

8.1 FORMULATION OF HYPOTHESIS

8.1.1 Hypothesis: What it is

A hypothesis may be defined as a proposition or a set of premises put forth as an explanation for the occurrence of some specified phenomenon. Hypotheses may be classified in terms of how they were derived, into, i) inductive, and ii) deductive. If a hypothesis involves generalization based on observation made it is known as inductive hypothesis. This is how general conclusions are drawn from particular facts or examples. Deductive hypothesis implies that it is derived from concepts or theory. A hypothesis is either asserted merely as a provisional conjecture to guide some investigation or accepted as probable in the light of established facts. Quite often a research hypothesis is a predictive statement, capable of being tested by scientific methods that relates an independent variable to some dependent variable.

Very often we make decisions about a population based on sample information from the population. Such decisions are known as "statistical decisions". While attempting to make such decisions, it is usual to make assumptions or guesses about the population involved. Such assumptions, which may or may not be true, are called "statistical hypotheses", and in general they are statements about the probability distribution of the population. A statistical hypothesis will be much more specific and amenable to statistical analysis. The statistical hypothesis consists of a null hypothesis, usually referred to as H_0, and an alternative hypothesis H_1 or H_a.

Experimentation provides a method of hypothesis testing. The researcher defines a problem and then proposes a hypothesis. With the help of test results, one can either confirm or reject the hypothesis. The confirmation or rejection is always stated in terms of probability rather than certainty. The ultimate purpose of experimentation is to generalize the variable relationship so that they may be applied outside the laboratory to a wider population of interest.

8.1.2 Null Hypothesis and Alternative Hypothesis

This is used in the context of statistical analysis. If we are to compare method A with method B about its superiority and if we proceed on the assumption that both the methods are equally good, then this assumption is termed as the "null hypothesis". As against this, we may think that the method A is superior or the method B is inferior, we are then stating what is termed as "alternative hypothesis". The null hypothesis is generally symbolized as H_0 and the alternative hypothesis as H_1. Suppose we want to test the hypothesis that the minimum yield strength for accepting a component made of steel is μ_0, then the null hypothesis can be symbolically expressed as,

$$H_{\ddot{u}} : \mu_0 = \mu \quad \text{or it could be simply written as}$$

$$H_0 : \mu = \mu_0$$

If our sample data does not support the null hypothesis, we conclude that something else is true. What we conclude by rejecting the null hypothesis is known as alternative hypothesis. In other words a set of alternatives to the null hypothesis is known as alternative hypothesis. We may consider three possible alternative as

   i. $H_1 : \mu \neq \mu_0$

   ii. $H_1 : \mu > \mu_0$

   iii. $H_1 : \mu < \mu_0$

Alternative hypothesis is usually the one which we wish to prove and the null hypothesis is the one which we wish to disprove or reject.

8.1.3 Level of Significance

It is an important concept in hypothesis testing. It is always indicated by some percentage (usually 5% or 1%), which should be chosen with great care, thought and reason. If we choose a 5% significance level, then this implies that H_0 will be rejected when the sampling result (observed evidence) has a less than 0.05 probability of occurring if H_0 is true. In other words, the researcher is willing to take as much as 5% risk of rejecting the null hypothesis when it (H_0) happens to be true. Thus, the level of significance is the maximum value of the probability of rejecting H_0 when it is true and is usually determined in advance before testing the hypothesis.

8.2 TESTING OF HYPOTHESIS

8.2.1 Type-I and Type-II Errors

While testing hypothesis, we are likely to make two types of errors. We may reject H_0 when H_0 is true or alternatively we may accept H_0 when in fact H_0 is false or not true. Rejecting H_0, when it is true is known as "Type-I error" and accepting H_0, when it is not true is called "Type-II error".

In other words type-I error means rejecting of a hypothesis which ought to have been accepted and type-II error means accepting the hypothesis which ought to have been rejected. Type-I error is denoted by 'α' (alpha) and known as α-error, also called the "level of significance of the test". The type-II error is denoted by β (beta) and known as β--error. These errors are shown in the decision tablegiven below.

State of nature	Decision	
	Accept H_0	Reject H_0
H0 (True)	Correct decision	Type-I Error (α)
H0 (False)	Type-II error (β)	Correct decision

If we fix type-I error at 5% (assuming level of significance as 5%), it means that there are about 5 chances in 100 that we will reject H_0, when H_0 is true. We can control type-I error by fixing it at a lower level. But with a fixed sample size of n, if we try to reduce type-I error, the probability of committing type-II error will increase. Both the types of errors cannot be controlled simultaneously. Hence, a trade-off between the two types of errors has to be made taking into account the cost or penalties attached to both types of errors.

In the example quoted in an earlier chapter 6, we conduct a test with ten specimens to get the mean value of yield strength for steel (μ_0) to be used in the construction of a building. The type-I error to reject H_0 when in fact it is true would then mean that steel would be shipped back to the supplier with resulting financial loss and penalty to the builder due to delay in construction. On the other hand, a type-II error, accepting H_0, when it is false, would mean that steel of inadequate strength would be used to construct the building, resulting in poor quality and perhaps failure of the building leading to loss. Thus, a type-II error should be minimized in engineering situations.

8.2.2 Two-Tailed and One-Tailed Tests

The terms two-tailed and one-tailed are quite significant in the context of hypothesis testing. The two-tailed test is appropriate when the null hypothesis, H_0, is some specified value and the alternative hypothesis, H_1, is a value not equal to the specified value, which means, it may be more or less than the specified. Therefore, the two-tailed test is an appropriate statistical test used in inference in which a given statistical hypothesis (H_0), will be rejected when the value of the statistic is either sufficiently small or sufficiently large. The test is named after the 'tail' of data under the far left or far right of a bell-shaped normal distribution curve. However, this terminology is extended to tests relating to distribution other than normal. Symbolically, the two-tailed test is appropriate when we have

$$H_0 : \mu = \mu_0 \text{ and } H_1 : \mu \neq \mu_0$$

which may be

$$\mu > \mu_0 \text{ or } \mu < \mu_0$$

Thus, in two-tailed test there are two rejection regions, one on each tail of the curve as shown in Figure 8.1.

Reject H_0 if the sample mean × falls in either of the region (two-taild test at 5%).

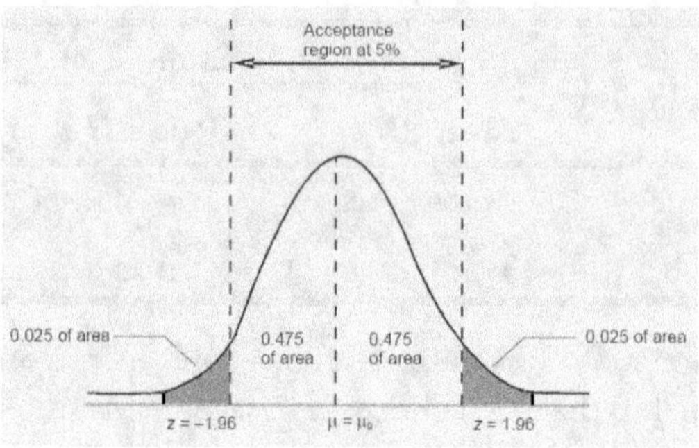

Figure 8.1 A two-tailed test

Reject if the sample mean × falls in either of the regions (one-taild test at 5%).

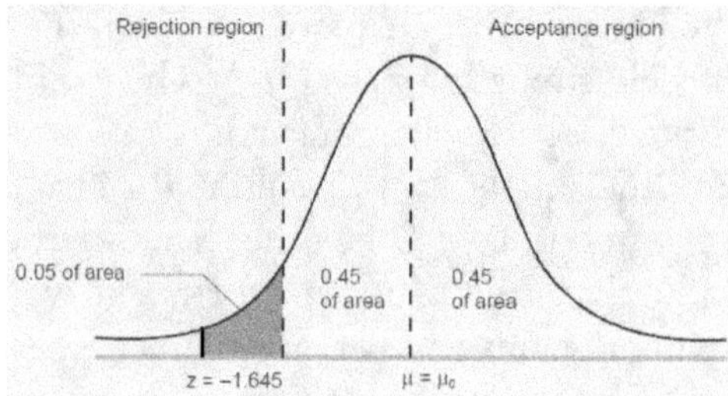

Figure 8.2 A one-tailed test

If the significance level is 5% and the two-tailed test is to be applied, the probability of the rejection area will be 0.05 (equally split on both tails)and the acceptance region will be 0.95 of the area as shown in Figure 8.1. If our sample mean deviates from μ_0 in either direction and fall in the rejection region then we shall reject the null hypothesis, but if the sample does not deviate significantly from μ_0, and lie in the acceptance region we accept the H_0.

One-tailed test is used when the population mean is either lower or higher than some hypothesized value. In one-tailed test, the null hypothesis is rejected only for values of the test statistics falling into one specified tail of its sampling distribution. For instance, if our $H_0 : \mu = \mu_0$ and $H_1 : \mu > \mu_0$, then we are interested in what is known as left-tailed test (wherein there is one rejection region on the left tail) as shown in Figure 8.2. In case our null hypothesis and , we are then interested in what is known as right-tailed test and the rejection region will be on the right tail of the curve.

Whether to use a two-tailed test or one-tailed test in a given problem, it depends upon the nature of the problem and the null hypothesis H_0 and alternative hypothesis H_a. If the alternative hypothesis has two alternatives, then a two-tailed test is the most appropriate and if it contains only one alternative, the one-tailed test is the most appropriate one. It should be remembered that accepting H_0 on the basis of sample information does not constitute the proof that H_0 is true. It only means that there is no statistical evidence to reject it, but we do not assert that H_0 is true.

8.2.3 How to Proceed with Testing of Hypotheses

The various steps involved in testing a hypothesis can be listed as follows:

i.   Making a formal statement of the null hypothesis H_0 and alternative hypothesis H_1.

ii. Selection of significance level, generally in practice; 10% or 5% or 1% level is adopted for the purpose.

iii. Determination of appropriate sampling distribution.

iv. Selection of random sample and computing the sample characteristic.

v.   Determining the critical region for rejecting the null hypothesis, H_0.

vi. Comparing the probability. If the calculated probability is value in the case $\alpha/2$ of one-tailed test and in the case of two-tailed tests, then reject the null hypothesis.

Standard textbooks on statistics do give details of type of hypothesis with conditions and what test statistic to be used with the formula and criteria for rejection of H_0.

The tests of hypothesis (to test significance) can be classified as i) Parametric tests or standard tests of hypothesis; and ii) Non-parametric tests or distribution free tests. The parametric tests usually assume certain properties of the parent population from which we draw samples. Assumptions like observations must come from normal population, sample size should be large, and assumptions about population parameters such as mean, variance, etc. must hold good before parametric tests can be used. In case, the researchers do not want to make such assumptions, they can use non-parametric methods to test hypothesis.

The important parametric tests are i) Z-test ii) t-test, iii) χ^2-test and iv) F-test. All these tests are based on the assumption of normality, i.e., the source of data is considered to be normally distributed. In some cases it may not be normally distributed, but still we use it on account of the fact that we mostly deal with samples and that the sampling distributions closely approach normal distribution.

The Z-test is used for judging the significance of several statistical measures, particularly the mean. For conducting Z-test to check sample mean with population mean the statistic to be used is z. This has to be calculated from the data available and compared with table value for the assumed level of significance, α. This approach is more often used by researchers.

The t-test is based on t-distribution and is appropriate to compare the significance of difference between the means of two samples in the case of small samples (i.e., $n <$ 30). The χ^2-test is based on chi-square distribution and is a parametric test used for comparing a sample variance to a theoretical population variance. F-test, which is based on F-distribution, is used to compare the variance of two independent samples. This test is also used in the context of analysis of variance (ANOVA) for judging significance of more than two samples.

Example   A manufacturer of switches assures a mean life of 2000 hours with a standard deviation of 200 hours based on their testing of a large number of samples over a period of time. Later on, the company made a small change in the spring. To determine whether this has changed the life of the switch, a sample of 100 switches were tested and it gave a sample mean as 1960 hours and standard deviation of 180. Based on these results, can the firm claim that there is a change in the life of the product?

The assured life of 2000 hours can be assumed to represent the population mean μ and 200 hours to represent the population standard deviation σ.

The null hypothesis, $H_0 : \mu_0 = 2000$

Alternative hypothesis, $H_1 : \mu \neq \mu_0$

Significance level is not given, and hence one can assume $(\alpha) = 5\%$ or .05

The sample size (n) = 100.

As the sample size >30, we can assume that the sample is drawn from a normal distribution. σ is also known, i.e., $\sigma = 200$.

As the alternative hypothesis H_1 is $\mu \neq \mu_0$, it may be more than or less than μ_0 and hence two-tailed test i the appropriate one to be used.

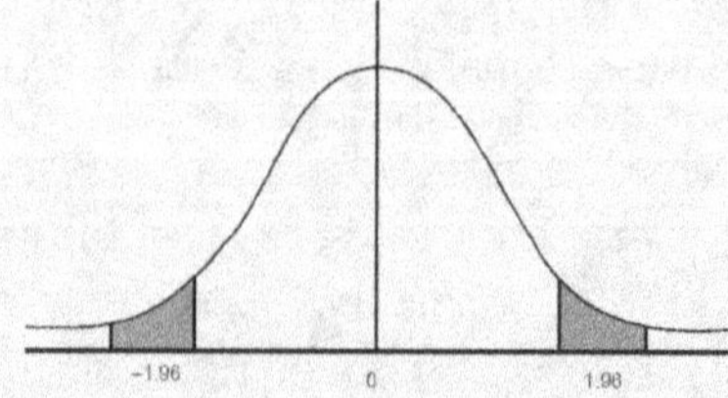

The critical region is and . From standard normal table, find values corresponding to $-z_{.025}$ and $z_{.025}$. The values obtained are -1.96 nd 1.96.

The calculated value of $z = \dfrac{\bar{x} - \mu}{\left(\sigma/\sqrt{n}\right)} = \dfrac{1960 - 2000}{180/\sqrt{100}} = \dfrac{-40}{18} = -2.22$

As the calculated value falls in the critical region, the null hypothesis is rejected and the firm can conclude that the change in spring has significantly altered the life of the switch. There are five chances in 100 to reject H_0, when it is true. One can minimize this error by selecting $\alpha = 1\% = .01$. Now the critical region is $z < -z_{.005}$ and $z > z_{.005}$. The corresponding table values are -2.58 and 2.58 respectively. Since the calculated z, (-2.22) falls outside the critical region, accept H_0 at 1% significance level.

In reality the firm is concerned only with whether the change in the spring reduced the life of the switch; it is not a serious consequence if the life was increased. Suppose, an alternative hypothesis, is formed as $H_1 = \mu < \mu_0$ that is $H_1 = \mu < \mu_0 = 2000$. In that case one has to deal with one-tailed test and the critical region for rejecting H_0 is $z < z_{-.01}$ or $z < -2.326$. Since $z = -2.22$, one can nt reject the null hypothesis, H_0.

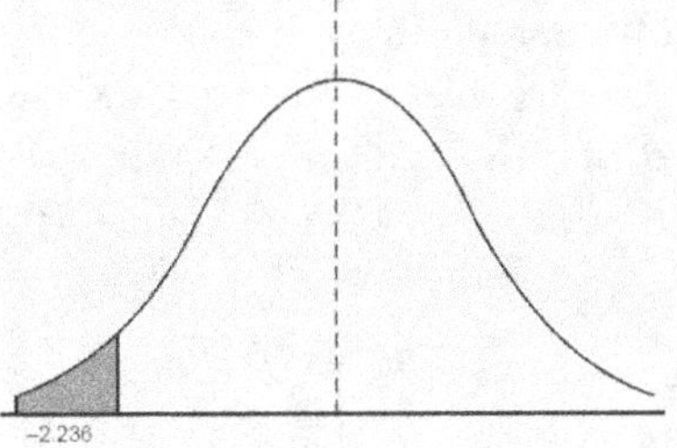

8.2.4 Limitations of Tests of Hypothesis

There are several limitations that a researcher should keep in mind while using tests of hypothesis. They include the following.

- It should not be used in a routine way. One should remember that test of hypothesis is not a decision-making tool by itself; but they are only aids to decision-making. Therefore, proper interpretation of statistical evidence is important for intelligent decision-making.

- The test results simply indicate whether the difference is on account of fluctuations of sampling or because of other reasons.

- When the test result shows that a difference is statistically significant, it only suggests that the difference is probably not due to chance.

- It cannot be said that statistical inferences based on the significance tests are entirely correct evidences concerning the truth of the hypothesis. The error in inferences is likely to be higher with small samples compared to larger samples.

8.3 ANALYSIS OF VARIANCE

8.3.1 What is ANOVA

Analysis of variance (ANOVA) is a statistical technique used when multiple sample cases are involved. The significance of difference between the means of two samples can be judged either through Z-test or t-test. But, when more than two samples are involved, the ANOVA technique enables us to, perform the significance test simultaneously. It is an important tool in the analyst's hand. Professor R.A. Fisher has made significant contribution by developing theory of ANOVA that has great use in practical applications. Later on Prof. Snedecor has greatly supplemented to its further development.

The essence of ANOVA is that the total amount of variation in a set of data is broken down into two types, the amount of variation which can be assigned to chance and the amount of variation which can be assigned to specific causes. There may be variation between samples and also within sample items. ANOVA comprises in splitting the variance for analytical purposes. Using ANOVA technique, one can investigate any number of input factors, which are said to influence the dependent variable. Here "factor" is used synonymously with "decision variable" or "input variable". In ANOVA one can investigate the differences between various categories within each of these factors which may have large number of possible values.

For example, the output of a process may be classified by machines and operators. From this cross classification, it could be determined whether the mean qualities of output of various operators differed significantly. Similarly, it could also be determined whether the mean qualities of outputs of various machines differ significantly. This study may help the researcher understand as to how to ensure uniformity in quality of output, whether by improving the operators or by standardization of machines.

We make two estimates of population variance, viz., one based on between-sample (treatment variance) and the other on within-sample (error variance) variance. The two estimates of population variance are compared using F-test by computing F-ratio as follows:

$$F = \frac{\text{Estimate of population variance between sample variance}}{\text{Estimate of population variance within sample variance}}$$

This value of F is then compared with the F-limit for a given degrees of freedom (from tables). If the calculated F-value is equal to or exceeds F-limit value, it is concluded that there is a significant difference between the samples or treatment means.

8.3.2 One-way Analysis of Variance

In one-way analysis, a single factor is considered at different levels (treatments) with each treatment having certain number of observations. Consider a single factor at m levels or treatments with each treatment containing n observations. Let the jth obserain of the ith treatment be denoted by x_{ij}.

Treatment	Observations	Total	Mean
1	$x_{11}\, x_{12} \ldots X_{1j} \ldots x_{1n}$	$X_{1.}$	$\bar{x}_{1.}$
2	$x_{21}\, x_{22} \ldots X_{2j} \ldots x_{2n}$	$X_{2.}$	$\bar{x}_{2.}$
i	$x_{i1}\, x_{i2} \ldots X_{ij} \ldots x_{in}$	$X_{i.}$	$\bar{x}_{i.}$
m	$x_{m1}\, x_{m2} \ldots X_{mj} \ldots x_{mn}$	$X_{m.}$	$\bar{x}_{m.}$

$$X_{i.} = \sum_{j=1}^{n} X_{ij}$$

$$\bar{\bar{u}}_i = \frac{\sum_{j=1}^{n} x_{ij}}{n} \; ; \text{ for } = 1, 2, \ldots$$

The "dot" subscript notation implies summation over the subscript it repaces. Therefore the grand mean $\bar{x}$ is given by,

$$\bar{x}... = \frac{\sum_{i=1}^{m} \sum_{j=1}^{n} X_{ij}}{m.n}$$

The total variability of the observations described by the total sum of squares (SST) is given by

$$\text{SST} = \sum_{i=1}^{m} \sum_{j=1}^{n} \left(X_{ij} - \bar{X}.. \right)^2$$

The total sum of squares can be partitioned into treatment sum of squares or sum of squares between samples (SSTr) and on error sum of squares (SSE) or within sum ofsquars. That is,

$$\text{SST} = \text{SSTr} + \text{SSE},$$

where,

$$\text{SST} = n \sum_{i=1}^{m} \left(X_{i.} - \bar{X}.. \right)^2$$

and

$$SST = \sum_{i=1}^{n} \sum_{j=1}^{m} \left(X_{ij} - \bar{X}.. \right)^2$$

The SSTr is based on the difference between the treatment means and the grand mean, while SSE is based on the difference between an observation within a treatment and the treatment mean. To get the mean square (MS) of SST, SSTr and SSE, divide each quantity by its appropriate degrees of freedom. For example, the d.f. for total variance is equal to the total number of items minus 1, i.e., $(m.n - 1)$. The d.f. for SSTr is $(m - 1)$ and for SSE it is $m(n - 1)$. Th one-way ANOVA table will be as given below.

Source of variation	Sum of squares	Degrees of freedom (d.f.)	Mean square (MS)	F-ratio
Between treatments (or) samples SSTr	$n\sum_{i=1}^{m}(\bar{X}_{i.}-\bar{X}..)^2$	$m - 1$	$\dfrac{SSTr}{(m-1)}$	$\dfrac{MS_{between}}{MS_{with}}$
Samples (SSTr) within the treatments (SSE)	$\sum_{i=1}^{m}\sum_{j=1}^{n}(X_{ij}-\bar{X}_{.})^2$	$m(n-1)$	$\dfrac{SSE}{m(n-1)}$	
Total sum of squares (SST)	$\sum_{i=1}^{m}\sum_{j=1}^{n}(X_{ij}-\bar{X}..)^2$	$(mn - 1)$	–	

To test the null hypothesis that the means of m treatments are equal, we have to compare the variance based on the variation between treatment means, i.e., MS for SSTr with the variance based on variation within the means, i.e., MS for SSE by computing F-ratio as shown in the last column of the above table. If the F-ratio calculated is larger than the F-statistic for $F_{\alpha, m-1, m(n\alpha-1)}$ taken from table for level of significance α, we reject the null hypothesis and conclude that the variable x is influenced by the level of treatments.

Example Suppose a factor is given three levels of treatment and for each treatment, there are four observations as given in the table, find whether the factor levels re significant.

Levels	Observations					Total	Mean
A	5	6	7	4	8	30	6
B	4	5	5	4	7	25	5
C	3	5	4	3	5	20	4

$$\sum \bar{x} = 75$$

$$\bar{x} = \frac{75}{15} = 5$$

Total sum of squares (SST) $= \sum\limits_{i=1}^{3}\sum\limits_{j=1}^{5}\left(x_{ij} - \bar{x}..\right)^2$

$$= \left[(5-5)^2 + (6-5)^2 + (7-5)^2 + (4-5)^2 + (8-5)^2\right]$$
$$+ \left[(4-5)^2 + (5-5)^2 + (5-5)^2 + (4-5)^2 + (7-5)^2\right]$$
$$+ \left[(3-5)^2 + (5-5)^2 + (4-5)^2 + (3-5)^2 + (5-5)^2\right]$$
$$= 15 + 6 + 9 = 30$$

Sum of squares within treatments (SSE) $= \sum\limits_{i=1}^{3}\sum\limits_{j=1}^{5}\left(\ddot{u}_{ij} - \bar{}_{i.}\right)^2$

$$= \left[(5-6)^2 + (6-6)^2 + (7-6)^2 + (4-6)^2 + (8-6)^2\right]$$
$$+ \left[(4-5)^2 + (5-5)^2 + (5-5)^2 + (4-5)^2 + (7-5)^2\right]$$
$$+ \left[(3-4)^2 + (5-4)^2 + (4-4)^2 + (3-4)^2 + (5-4)^2\right]$$
$$= 10 + 6 + 4 = 20$$

Sum of squares between treatments (SSTr) $= 5 \sum\limits_{i=1}^{3}\left(x_{i.} - \bar{x}..\right)^2$

$$= 5\left[(6-5)^2 + (5-5)^2 + (4-5)^2\right] = 10$$

Source of variation	SS	d.f	MS	F-ratio	$\alpha = 5\%$ F-limit
Between treatments	10	3 – 1 = 2	10/2 = 5	5/1.67 = 2.99	$F_{.05,\ 2,\ 12}$ = 3.88 (table)
Within treatments	20	3(5 – 1) = 12	20/12 = 1.67		
Total	30	(3 × 5) – 1 = 14			

The calculated value of *F* is 2.99 which is less than the table value of 3.88 at 5% level of significance with d.f. v_1 = 2, v_2 = 12. Hence, the null hypothesis of no difference between means could not be rejected. The difference may be due to chance.

8.3.3 Two-way Analysis of Variance

Two-way ANOVA technique is used when the data is classified on the basis of two factors. For instance in the case of process output, the two factors considered could be operators and machines. Such a two-way design may have repeated measurements of each factor or may not have repeated values. The technique is little different in the case of repeated measurements where we also compute interaction variations. The arrangement of data table for bothwith and ithout replication will be as follows:

Treatment	Factors	
	F_1	F_2
T_1	xxx	xxx
T_2	xxx	xxx
T_3	xxx	xxx
T_4	xxx	xxx
T_5	xxx	xxx

Treatment	Factors	
	F_1	F_2
T_1	xxx	xxx
	xxx	xxx
	xxx	xxx
	xxx	xxx
	xxx	xxx
T_2	xxx	xxx
	xxx	xxx
	xxx	xxx
	xxx	xxx
	xxx	xxx

i. **When repeated values are not there** As we do not have repeated values, we cannot directly compute the sum of squares within treatments (SSE) as we had done in the case of one-way analysis. We have to calculate this residual or error variation by subtraction, once we have calculated the sum of squares for total variance (SST) and for the variance between treatments (SSTr), i.e., SSE = (SST − SSTr − SSB), where SSB is the sum of square between columns (blcks). The total sum of squares (SST) is given by

$$\text{SST} = \sum_{i=1}^{m} \sum_{j=1}^{n} \left(x_{ij} - \bar{x}.. \right)^2 \quad \text{d.f.} : \left(m.n - 1 \right)$$

$$\text{SSTr} = n \sum_{i=1}^{m} \left(x_i - \bar{x}.. \right)^2 \quad \text{d.f.} : \left(m - 1 \right)$$

$$\text{SSB} = m \sum_{j=1}^{n} \left(x_{.j} - \bar{x}.. \right)^2 \quad \text{d.f.} : \left(n - 1 \right)$$

$$\text{SSE} = \sum_{i=1}^{m} \sum_{j=0}^{n} \left[\left(x_{ij} - \bar{x}_{i.} \right) \left(x_{.j} - \bar{x}.. \right) \right]^2 \quad \text{d.f.} : \left(n - 1 \right)\left(m - 1 \right)$$

where,

m = number of treatments (rows) and

n = number of columns (blocks)

The ANOVA table can be set up as shown.

The two-way ANOVA table can be set up as given below. The mean square (MS) residual or residual variance provides the basis for calculating F-ratios. The MS residual is always due to the fluctuations in sampling and hence serves as the basis for the significance test. Both the F-ratios are compared with their corresponding table values for a given df and at a specified level of significance. As usual if it is found that the calculated F-ratio concerning variation between columns (blocks) is equal to or greater than the critical value, then the difference between block means is considered significant. Similarly, the F-ratio concerning variatio between treatment means (rows) can be interpreted.

Source of variation	Sum of squares	d.f.	Mean square (MS)	F-ratio
Between block (columns) SSB	$m\sum\limits_{j=1}^{n}\left(x_{.j}-\bar{x}..\right)^2$	$(n-1)$	$\dfrac{SSB}{(n-1)}$	$\dfrac{MS\ of\ SSB}{MS\ of\ SSE}$
Between treatment (rows) SSTr	$n\sum\limits_{i=1}^{m}\left(x_{i.}-\bar{x}..\right)^2$	$(m-1)$	$\dfrac{SSTr}{(m-1)}$	$\dfrac{MS\ of\ SSTr}{MS\ of\ SSE}$
Residual error SSE	SST – SSTr– SSB	$(n-1)(m-1)$	$\dfrac{ü}{(n-1)(m-1)}$	
Total SST	$\sum\limits_{i=1}^{m}\sum\limits_{j=1}^{n}\left[\left(x_{ij}-\bar{x}..\right)\right]^2$	$(m.n-1)$		

ii. *When repeated values are there*   In the case of two-way design with replication we can obtain a separate independent measure of inherent or smallest variations. By introducing replication, we are able to separate out the variation due to interaction effects from that due to random error. An interaction effect is present if the observed values for different levels of one factor are altered by the presence of the other factor. The sum of squares due to error within each cell (sample for each treatment) is determined by first determining the subtotal sum of squares (SSS) and then sum of square within cells (SSWC) computed by subtracting SSS from SST, i.e., SSWC = (SST – SSS). Finally the interaction sum of squares is calculated as,

$$SSI = [SSS - (SSTr + SSB)]$$

Suppose, the number of columns (blocks) is n, and the number of treatments m, and within ech treatment the number of replications is r, then,

$$SST = \sum x_y^2 - \frac{T..^2}{m.n.r}$$

where, T is the cell subtotal, and $T..$ is the grand toal.

The treatment sum of squares (between rows) is,

$$SST = \sum \frac{Ti^2}{n.r} - \frac{T..^2}{m.n.r}$$

where, $T_{i.}$ is the sum of column total of cell toals.

The block sum of squares (between columns) is,

$$SSB = \sum \frac{Ti^2}{m\,ür} - \frac{ü^2}{m\ \ n\ \ r}$$

where, $T_{.j}$ is the sum of each column

$$SSB = \sum \frac{T^2}{r} - \frac{T..^2}{m.n.r}$$

Sum of squares within cells,

$$SSWc = [SST - SSS]$$

The interaction sum of squares is given by

$$SSI = SSS - SSTr - SSB$$

Once the computations are made, the resuls could be tabulated in the ANOVA table as follows.

Source of variation	SS	d.f.	MS	F-ratio	5% F-limit
Between blocks (columns)	SSB	(n – 1)	$\dfrac{SSB}{(n-1)}$	$\dfrac{MS\ of\ ü}{MS\ of\ SSWC}$	
Between treatments	SSTr	(m – 1)	$\dfrac{SSTr}{(m-1)}$	$\dfrac{MS\ of\ SSTr}{MS\ of\ SSWC}$	
Within cells (error)	SSWC	m. n (r – 1)	$\dfrac{SSWc}{m.n(r-1)}$	–	
Interaction	SSI	(m – 1) (n – 1)	$\dfrac{ü}{(n-1)(m-1)}$	$\dfrac{MS\ of\ SSI}{MS\ of\ SSWC}$	
Total	SST	(m. n. r – 1)	–		

To check whether there is a significant interaction between the two factors, the F-ratio forinteraction is checked against the tabulated value.

9

Design of experiments and regression analysis

9.1 DESIGN OF EXPERIMENTS

9.1.1 Importance of Experimental Design

Experimental studies generally provide a logical and systematic approach to answer questions, such as "what will happen if certain variables are carefully controlled or manipulated?" In fact deliberate manipulation of input factors is a part of the experimental method as it helps the researcher to measure the effect of the factor manipulated intentionally. Experimentation not only provides a method for hypothesis testing but also permits drawing of inferences about causality. As a result of experiments, the researcher is able to arrive at significant conclusions and also discern the science behind the phenomena.

When experiments are planned in advance using statistical methods, it helps to gain greatest benefit. Proper planning ensures that data are collected in a way that will provide the most unbiased and precise results commensurate with the desired expenditure and time spent. Major benefits from statistically designed experiment are,

- More information per experiment will be obtained than with unplanned way of conducting experiments.
- Provides an organized approach to the collection and analysis of data.
- Credibility of the conclusions of experimental programme is enhanced since the variability and sources of experimental error are analysed by using appropriate statistical methods.
- Ability to establish interactions between experimental variables are possible.
- Due to the emphasis on improving quality nowadays, the experimental design approaches are assuming greater significance.
- Use of techniques based on orthogonal arrays for experimental design is not only unique but also helps to produce robust results.

9.1.2 Randomization

If an experimental programme involves large number of tests, it is necessary to randomize the order in which the specimens or combination of factors are selected for testing. Thus, by randomization each one of the specimen involved in the experiment has an equal chance of being selected for a given test. By this approach, bias due to uncontrolled second-order (extraneous) variables is minimized. For example, in any extended testing programme, errors can arise over a period of time owing to subtle changes in the characteristics of testing equipment or in the proficiency of the operator of the test. Similarly, when test specimens are prepared from large forgings or ingots, the possibility of variation of properties in the forgings must be considered. By randomly selecting places to cut specimen from a large forging the variation between specimen could be minimized.

9.1.3 Variables and Factors

Any experimental investigation starts with preliminary familiarization of experiments. This will ensure better understanding and defining the problem. Once the goals of investigation are well-defined, the large number of variables which are possible should be reduced to a few important ones. The statistical methods will be of much use in this. As the experiment moves to optimization stage, the number of variables can be reduced further. Here, again, statistical design serves effective means.

The output of an experimental run is the data termed as response variable(s). Response variables can be classified as quantitative, qualitative or quantal. The quantitative response, which is measured by a continuous scale, is the most common and easiest to work with in statistical analysis. Qualitative responses like lustre, odour, etc. can be ranked on an ordinal scale, such as worst (0) to best (1). Quantal or binary response produces one or two values "go" or "no-go", i.e., pass or fail.

Factors are experimental variables that are supposed to be controlled by an experimenter. An important part of planning of an experimental program is identifying the significant variables that affect the response and deciding how to exploit them in the experiment. Factors may be independent in the sense that the level of the effect of one factor is independent of the effect of level of other factors. However, two or more factors may have interacting influence with one another, i.e., the effect on the response of one variable depends upon the levels of other variables.

In addition to primary variables that are under the control of the experimenter, there are other variables which may not be strictly under the experimenter's control. The use of randomization of test runs removes this unconscious bias. Also, the use of experimental blocks produces relatively homogeneous test conditions and minimizes this. When an experiment is conducted using blocking concepts, the effect of background variables is removed from the experimental error to a greater extent.

9.2 PLANNING OF EXPERIMENTS

9.2.1 Classification of Experimental Designs

One should realize that not all primary factors are capable of variation with equal facility. Complete randomization may not also be practicable. Often the final experiment is a compromise between information that can be obtained and the cost involved in that. Therefore, while developing the experimental design, the physical reality gets precedence over strict adherence to statistical nuances.

It is always better to make some initial assessment of the overall repeatability before embarking on major experimental test programme. Experience of previous experiments may also be useful in this. In case pilot experiments show large variability in the response, it indicates that primary variables have not been properly identified. It is not necessary to conduct statistically designed experiments all the way through completion. In fact there are benefits in conducting experiments in stages, taking into account the advantage of information gathered from results of earlier stage of the experiment.

The statistically designed experiments may be classified as given below.

i. *Blocking designs*   The blocking designs by nature use blocking concepts to remove the effect of background variables from the experimental error. Common type of blocking designs include,

Randomized block and	:	Removes the effect of single
balanced incomplete block	:	extraneous variables
Latin square and Youden square	:	Removes the effect of two extraneous variables
Graeco-Latin square or and hyper-Latin square	:	Remove the effect of three more extraneous variables

ii. *Factorial designs*   Factorial designs are experiments in which the levels of each factor are combined with the levels of all other factors to conduct experiment simultaneously. This is an important class of experiment for an analyst. For example, if there are three factors with each factor having four levels, then there will be 64 combinations (4^3) of experiments to be conducted. In factorial designs we have a) Full factorial designs b) Fractional factorial design and c) Use of orthogonal arrays.

iii. *Response surface designs*   The response surface designs are used to develop empirical functional relation between the factors (independent variables) and the response (output variables).

9.2.2 Full Factorial Design

A full factorial design is one in which we control several factors and investigate their effects at each of two or more levels. The experimental design consists of making an observation at each of all possible combinations that can be formed for the different levels of factors. Each different combination is called a "treatment combination". The approach adopted in a factorial experiment is different compared to traditional experiment wherein all factors but one are held constant in each trial. The simplest and most common type of factorial design is one that uses only two levels, i.e., 2n factorial design, where n represents the number of factors.

To illustrate the point, let us consider three factors A, B, C each at two levels, one at low level indicated by a (–) sign and the other at higher level indicated by a (+) sign. We are interested in the result of the experiment due to change in factor A alone, B alone and C alone. These three potential output changes are called "main effects". The interaction effect between factors shows whether the effect of one factor depends on the levels of the other factors. There are first-order effects due to joint change in factors A and B, A and C and B and C. These interactions are denoted by AB, AC and BC respectively. Likewise, the second-order interaction effect due to simultaneous changes in all the three factors A, B and C is denoted by ABC. This experiment is known as three-factor 2^3 designs.

In 2^3 factorial design, the experimental design matrix willbe as given below:

Design points	Levels of factors		
	A	B	C
1	–	–	–
2	+	–	–
3	–	+	–
4	+	+	–
5	–	–	+
6	+	–	+
7	–	+	+
8	+	+	+

Yate has introduced a unique notation, and adopting this notation, the three factor 2^3 factorial design table willbe as shown below:

Yate's std. order	Expt.No.	Level of factors							Response
		X_A	X_B	X_C	$X_A X_B$	$X_A X_C$	$X_B X_C$	X_{ABC}	
1	1	–	–	–	+	+	+	–	Y_1
a	2	+	–	–	–	–	+	+	Y_a
b	3	–	+	–	–	+	–	+	Y_b
ab*	4	+	+	–	+	–	–	–	Y_{ab}
c	5	–	–	+	+	–	–	+	Y_c
ac*	6	+	–	+	–	+	–	–	Y_{ac}
bc*	7	–	+	+	–	–	+	–	Y_{bc}
abc@	8	+	+	+	+	+	+	+	Y_{abc}

* : first-order interaction, @ : second-order interaction

The responses of these factor levels are denoted by notations Y_1, Y_a, Y_b, Y_c, Y_{ab}, Y_{ac}, Y_{bc} and Y_{abc}. The main effect of a factor is the average change in the response produced by a change in the level, say from its low level (–) to the high level (+), holding all other factors fixed at the same time. For example, the main effect of factor 'j, denoted by X_j is

$$X_j = \frac{\left(\text{Sum of responses at high } X_j\right) - \left(\text{Sum of responses at low } X_j\right)}{\text{Half the number of experiments}}$$

In terms of Yate's notation, the main efect of factor A is,

$$X_A = \frac{\left(Y_a + Y_{ab} + Y_{ac} + Y_{abc}\right) - \left(Y_1 + Y_b + Y_c + Y_{abc}\right)}{4}$$

An analyst may also be interested in finding the interaction effect between factors. For example, the effect of interaction betweenfactors B and C is

$$X_{BC} = \frac{\left(Y + Y + Y + Y\right) - \left(Y + Y + Y + Y\right)}{4}$$

The approach stated above can be illustrated by means of a hypothetical numerical example.

Example A certain variety of steel is subjected to three treatments, A, B, C by setting each treatment parameter at two levels, viz., a low level and, a high level. The strength measured in MPa for the 2^3 factorial experiment is givenin the table below:

Expt. No.	Factor levels			Response (MPa)
	A	B	C	
1	–	–	–	1041
2	+	–	–	1014
3	–	+	–	1014
4	+	+	–	696
5	–	–	+	1255
6	+	–	+	1096
7	–	+	+	1193
8	+	+	+	1089

i. Find the main effects of factors A, B and C and also the interaction effects.

ii. Check whether there is any significance between effect of levels of factor A by conducting appropriate statistical tests.

Using the Yate's notation for conducting 2^3 factorial design, the response table for the combination of factor levelswill be as follows:

Treatment	Levels of factors							Response (MPa)
	X_A	X_B	X_C	X_{AB}	X_{AC}	X_{BC}	X_{ABC}	
(1)	–	–	–	+	+	+	–	1041
a	+	–	–	–	–	+	+	1014
b	–	+	–	–	+	–	+	1014
ab	+	+	–	+	–	–	–	696
c	–	–	+	+	–	–	+	1255
ac	+	–	+	–	+	–	–	1096
bc	–	+	+	–	–	+	–	1193
abc	+	+	+	+	+	+	+	1089

The main effect of factor 'A' is calculated as indicated in the prvious section,i.e.,

$$X_A = \frac{1}{4}\left[(1014 + 696 + 1096 + 1089) - (1041 + 1014 + 1255 + 1193)\right] = -152.0$$

For factor B

$$X_B = \frac{1}{4}\left[(1014 + 696 + 1193 + 1089) - (1041 + 1014 + 1255 + 1096)\right] = -103.5$$

For factor C

$$X_C = \frac{1}{4}\left[(1255 + 1096 + 1193 + 1089) - (1041 + 1014 + 1014 + 696)\right] = 217$$

The minus sign for factors A and B indicates that their lower levels have contributed for the higher response. However, in respect of factor C, the higher level of input has contributed to higher response.

Similarly, the interaction effects can also be compute as indicated below:

$$X_{AB} = \frac{1}{4}\left[(1041 + 696 + 1255 + 1089) - (1014 + 1014 + 1096 + 1193)\right] = -59.0$$

$$X_{AC} = \frac{1}{4}\left[(1041 + 1014 + 1096 + 1089) - (1014 + 696 + 1255 + 1193)\right] = -20.5$$

$$X_{BC} = \frac{1}{4}\left[(1041 + 1014 + 1193 + 1089) - (1014 + 696 + 1255 + 1096)\right] = 69.0$$

$$X_{ABC} = \frac{1}{4}\left[(1014 + 1014 + 1255 + 1089) - (1041 + 696 + 1096 + 1193)\right] = 86.5$$

To compare the mean value of factor A at its high level with the mean value at low level, a paired t-test can be conductd sdtailed below:

$$\bar{X}_L = 1126$$

$$\bar{X}_H = 974$$

$$\bar{d} = \bar{X}_L - \bar{X}_H = 1126 - 974 = 152$$

Also

$$\bar{d} = \frac{\sum d}{4} = \frac{608}{4} = 152$$

$$\bar{X}_L = 1126, \ \bar{X}_H = 974, \ \sum d = 608$$

Low value A	High value A	Difference d	Deviation from d
1041	1014	27	−125
1014	696	318	166
1255	1096	159	7
1193	1089	104	−48

$$s^2 = \frac{\sum (deviation)^2}{(n-1)} = \frac{(-125)^2 + (-166)^2 + (7)^2 + (-48)^2}{(4-1)}$$

$$s^2 = 15178; \qquad s = \sqrt{15178} = 123.2$$

Assuming a null hypothesis that there is no difference in the mean values between its high and lo level for factor A,

i.e., $\mu_L = \mu_H$; $= \mu_L - \mu_H = 0$

$H_0 : \mu = 0$ and $H_1 : \mu \neq 0$

the t-statistic an be obtained from

$$t = \frac{\bar{d}}{s/\sqrt{n}} = \frac{\ddot{u}}{123.2/\sqrt{4}} = 2.47$$

From students t-distribution table, for two-tailed test, $t_{3,0.1} = 2.35$. As the calculated value of t is greater than table value the null hypothesis is rejected at a 10% level of significance. This implies that there is significant difference in the mean values between the high level and low level of treatment for factor A. Similarly paired t-test can be used to check the significance of interaction effects. The precision of statistical analysis can be improved by replication.

9.2.3 Fractional Factorial Design

Because of cost and time considerations, an analyst may wish to investigate an experimental model with less than full factorial design. This type of partial layout is called fractional factorial design and it is usually used when four or more input factors are considered simultaneously. The number of treatment combinations in a factorial design increases with increase in the number of factors. For example if $n = 3$, with 2 levels, the number of combinations will be, $2^3 = 8$ and if $n = 5$, it will be $2^5 = 32$, but if $n = 8$, it will be $2^8 = 256$ and so on. If the number of main effects will be n, the first-orde interaction will be $\frac{n(n-1)}{2}$, second-oder interaction will be $\frac{n(n-1)(n-1)}{(2)(3)}$ and so on. For example, in the case of 2^5 factorial experiment, there will be 5 main effects: since n = 5;

First-order interaction	:	$n(n-1)/2 = 10$,
Second-order interaction	:	$n(n-1)(n-2)/(2)(3) = 10$;
Third-order interaction	:	$n(n-1)(n-2)(n-3)/(2)(3)(4) = 5$:
Fourth-order interaction	:	$n(n-1)(n-2)(n-3)(n-4)/(2)(3)(4)(5) = 1$;

By neglecting higher order interaction effects, one can design an experiment involving some fraction of the full factorial design. For example, if he/she considers one half replicate of 2^3 factorial design, then $N = r\,(2^n) = \frac{1}{2}\,(2^3) = 4$. Therefore by splitting the treatment contributions into two blocks the main effects of A, B, and C can be studied by using fractional factorial design. This enables a researcher to conduct a series of short experiments which permit him/her to rapidly screen many variables for their effect and then select a few important ones.

9.3 EXPERIMENTS USING TAGUCHI APPROACH

9.3.1 Orthogonal Matrix Experiments

Full factorial design of experiments is conducted under all combinations of factor levels. For example, if an experiment consists of three factors at four levels each, there is a need to conduct $4^3 = 64$ experiments to try all combinations. Of course, the major advantage is that interaction effects could also be studied in this. Fortunately in many practical situations, the interaction effect is significantly less or absent and hence additive models provide an excellent approximation. In this context, employing the concept of orthogonal arrays serves as a powerful tool for economical exploration of relations that involve many factors.

Designed experiments can be considered as matrix experiments. It consists of a set of experiments where we change the settings of various input factors from one experiment to another. Orthogonal arrays are special matrices which permit the effect of several factors to be studied efficiently. An array used in matrix experiments is called an orthogonal array if all its columns are mutually orthogonal.

Here, the orthogonality implies that for any pair of columns, all combination of factor levels occurs and that they occur equal number of times. This is called the balancing property. It has been established that the balancing property is a sufficient condition for the matrix experiment to be orthogonal. In order to preserve the benefit of using an orthogonal array, it is imperative that all experiments in the matrix are performed. If experiments corresponding to one or more rows are not conducted, or if their data are missing or erroneous, the balancing property and, hence the orthogonality is lost.

9.3.2 Taguchi's Approach

Dr. Genichi Taguchi developed a method based on orthogonal array for designing experiments to investigate how different factors affect the response especially the mean and variance. His approach gave much reduced variance for the experiment optimum setting of control factors. In fact, Taguchi has coupled design of experiments with optimization of control factors to ensure best results. He has also introduced the concept of signal-to-noise ratio (S/N), where the signal is the mean response and noise is variation due to uncontrolled factors. These signal-to-noise ratios are nothing but log functions of desired response and also serve as objective function for optimization in experimental data analysis.

In Taguchi's method instead of testing all possible combinations like full factorial design, pairs of combination of factor levels are used to determine which factor affect mostly with minimum number of experiments. Taguchi has tabulated 18 basic orthogonal arrays called "Standard orthogonal arrays", which are available in many standard texts. The process of fitting an orthogonal array to a specific experimental study has been made easy by employing one of the standard orthogonal arrays. An example of Taguchi's orthogonal array, $L9(3^4)$ is given below:

The array $L9(3^4)$ has 9 rows and four 3 level columns. Similarly array $L18(2^1 3^7)$ has 18 rows, one 2 level column and seven 3 level columns. The number of rows of an array represents the number of experiments to be conducted. For any pair of columns all combination of factor levels occurs and they occur equal number of times. In the $L9(3^4)$ array for each pair of column, there are 9 combinations of factor levels and such combination precisely occurs only once. For example, if we take column 1 and 2 of L9 array, the combination of factor levels are: (1,1); (1,2); (1,3); (2,1); (2,2); (2,3); (3,1); (3,2); (3,3), thus complying the balancng property of orthogonal array.

Expt. No.	Columns			
	1	2	3	4
1	1	1	1	1
2	1	2	2	2
3	1	3	3	3
4	2	1	2	3
5	2	2	3	1
6	2	3	1	2
7	3	1	3	2
8	3	2	1	3
9	3	3	2	1

L9 (3^4) orthogonal array

9.3.3 Selection of Suitable Array

The experimenter after deciding the number of input factors and the number of levels of each factor, can select the proper orthogonal array for his/her experiment from the selection table. A portion of the orthogonal array selector tale is given below for reference.

Number of levels	Number of factors								
	2	3	4	5	6	7	8	9	10
2	L4	L4	L8	L8	L8	L8	L12	L12	L12
3	L9	L9	L9	L18	L18	L18	L18	L27	L27
4	L'16	L'16	L'16	L'16	L'32	L'32	L'32	L'32	L'32
5	L25	L25	L25	L25	L25	L50	L50	L50	L50

The name of the appropriate array can be found by looking at the column and row corresponding to the number of input factors and the number of levels considered. Suppose, one considers three factors such as cement, aggregate size and water–cement ratio, each factor at 3 levels, then the appropriate orthogonal array is L9. Once the name of the array is selected by looking at the concerned pre-defined array, the number of experiments and the combination of levels of each factor can be found out. In this example, the concerned pre-defined array is $L9(3^4)$, where 9 refers to the number of experiments to be conducted, 3 refers to the number of levels and 4 refers to the input factors.

Alternately one can decide by determining the total degrees of freedom. The degrees of freedom provide the clue on minimum number of experiments to be performed to study all chosen factors. Suppose in the example considered above, the total number of d.f. is 9, i.e., $4(3 - 1) + 1 = 8 + 1 = 9$. This is for 4 factors at 3 levels each and one is added for the overall mean. This indicates that a minimum of 9 experiments is required.

In certain cases, the number of levels for each factor may not be equal. For instance, in a study there are four factors A, B, C and D and levels are not equal as shown below:

Factors	A	B	C	D
No. of levels	3	3	3	2

In this case also, one can select an L9 array, by noting that the mean of levels is approximately equal to 3, i.e., $\frac{1}{4}(3 + 3 + 3 + 2) = 11/4$. Even if the total d.f. is calculated, it is equal to 8, i.e., $((2 + 2 + 2 + 1) + 1 = 8)$ which indicates the minimum number of experiments to be performed. But, however, as the average number of levels is 3 and the number of factors is 4, selection of $L9(3^4)$ is appropriate.

9.3.4 Experimental Steps

The steps involved in Taguchi's method of design of experiments are as follows:

i. Considering the objective and the nature of experiment involved, the performance measure has to be identified. The objective may be maximization or minimization of the performance measure. For example, it may be maximization of strength of concrete based on the ingredients such as cement, aggregate and water–cement ratio. The deviation in performance characteristic from the target value is used to define loss function for the process.

ii. Identify the factors that affect the response of interest, by controlling which the response of the system can be maximized or minimized. The number of levels that the factors should be varied must be specified. For instance the water–cement ratio may be fixed as 0.3, 0.4 and 0.5.

iii. Based on the number of factors and the levels of each factor, select the proper orthogonal array to be used for the experiment. The discussion in the previous sections and the array selection table will be useful in this. The array selection assumes that each parameter has the same number of levels. Sometimes, it may not be true. Suppose there are 4 factors, 3 factors having 3 levels and the fourth one has only 2 levels. As the mean number of levels is 3, we have to use a value of 3 for choosing our orthogonal array. There are only 2 levels in the last factor and hence the last column of the experimental table has to be filled by randomly choosing the two levels in a balanced way.

iv. Perform the experiments as per the orthogonal array table and collect the data on the effect of the performance measure. If each experiment is repeated (number of trials conducted) then performance measure for each trial has to be gathered.

9.3.5 Analysis of Experimental Data

After gathering the experimental results of performance measure for each trial, the relative effect of different factors can be analysed. Taguchi used the concept of signal-to-noise ratio, a terminology from electrical engineering, and introduced it in a specific way to control the performance characteristics. He has combined both mean and variance by a single measure and called it as signal-to-noise ratio (S/N ratio). The signal is the mean of the response and the noise represents the variance. It depends upon the nature of objective of the experiment. Taguchi also classified the problem as static and dynamic. In the static problem, the performance to be optimized has several factors which directly decide the target or desired value. The optimization then deals with that combination of factors which gives the best results. He has introduced the following signal-to-noise ratios for optimization of static problems.

i. For nominal-the-best typeof situation, he proposes,

$$S/N = 10 \log \left[\bar{y}_i^2 \middle/ s_i^2 \right]$$

when $\bar{y}_i$ is the mean of all y_i and s_i^2 is the variance. While the mean of each experiment represents the signal, the variance represents the noise.

ii. For the case of minimizing the performance characteristic, (i.e., smaller-the-bette type) the S/N ratio to be used is,

$$S/N = -10 \log \left[\frac{1}{N} \sum y_i^2 \right]$$

iii. For the case of maximizing the performance characteristic (i.e., larger-the-bettr type) the S/N ratio to be used is,

$$S/N = -10 \log \left[\frac{1}{N} \sum \frac{1}{y_i^2} \right]$$

The dynamic problem envisages a signal input which directly decides the response or output and are controlled by certain control factors. The optimization involves deciding the control factor levels which makes the input signal/output ratio closest to the desired one.

After calculating the ratio for each experiment, the average S/N value is calculated for each factor and combination of levels. For example, in the case of L9 orthogonal array experiment, i corresponding to each run of experiment, the signal-to-noise ratio (R_i) is calculated using appropriate formula taking into account the objective of the experiment. Let us call tem R_1, R_2...R_9, as shown in the table.

Experiment No.	Factors				S/N ratio
	A	B	C	D	
1	1	1	1	1	R_1
2	1	2	2	2	R_2
3	1	3	3	3	R_3
4	2	1	3	3	R_4
5	2	2	1	1	R_5
6	2	3	2	2	R_6
7	3	1	2	2	R_7
8	3	2	3	3	R_8
9	3	3	1	1	R_9

For example from the S/N ratios of each experiment, the average value of S/N ratio for each factor and at a particular level is calculated as detailed below. For instance in the case of factor A with level 1, the S/N ratio is calculated as

$$R_{A1} = 1/3 \ (R_1 + R_2 + R_3)$$

Similarly for B_2 and C_3 can be calculated as,

$$R_{B2} = 1/3 \ (R_2 + R_5 + R_8)$$

$$R_{C3} = 1/3 \ (R_3 + R_5 + R_7)$$

The average S/N ratio for each factor and level calculated can be tabulated as shown below. From the table, keeping in mind the objective of the experiment, viz., maximization or minimization of the performance measure, the best leel for each factor can be selected.

Factor	Levels		
	1	2x	3
A	x	x	x
B	x	x	x
C	x	x	x
D	x	x	x

Taguchi method of experimental design also allows for the use of a noise matrix by including external factors affecting the process outcome rather than repeated trials. However, it increases the complexity of experiments.

9.3.6 Advantages and Disadvantages

Taguchi method is a powerful yet a simple tool and easy to apply to many engineering problems. It can be used to quickly narrow down the scope of the study or to identify problems in a process from the data already in existence. It permits the analysis of several input factors with less number of experiments.

The statistical analysis approach traditionally employs the "means" of responses for analysis, while Taguchi method uses signal-to-noise ratio S/N involving both mean and variability of response. Depending upon the nature of the problem, a suitable objective function proposed by Taguchi can be used. However, the main disadvantage of Taguchi method is that the results obtained are only relative and do not exactly indicate which factor has the highest effect on the performance characteristic value. It is not possible to establish relationship with all variables if needed.

Case Study
A researcher wanted to find an optimum combination of factors which influence the cutting force in a metal cutting process. Though there are several influencing factors such as cutting speed, feed, depth of cut, coolant used, tool geometry, tool material, etc. which affect cutting conditions, the following factors, viz., cutting speed, feed and cutting edge radius were considered as most prominent ones and included as input factors for the study. The response of interest is the major cutting force F_c developed. For each of the three input factors, three levels of factor settings were fixed. This gives rise to 27 combination of experimental design points ($3^3 = 27$) under full factorial design. The factors and levels choen are given in the following table.

Factor levels	Cutting speed in m/ minutes (A)	feed in mm/rev (B)	Cutting edge radius in µm/c (C)
1	100	0.1	0
2	200	0.3	100
3	300	0.5	200

As the full factorial design involves 27 experiments to be performed, the researcher decided to use Taguchi's orthogonal array approach to this problem. The selection of this approach is primarily on account of minimization of effort needed and also the reasonable level of confidence that ensues by following this approach. Since the proposed experiment involves 3 factors with each at 3 levels, total number of d.f. is 7 (3(3 – 1) + 1) and therefore it calls for 7 minimum number of experiments. As the number of factors are 3 with each at 3 levels the standard array L9(3^3) was selected for the study using the array selection table. This requires 9 experiments to be conducted with factor combinations as stipulated in L9(3^3) arrayas indicated in the following table.

Experiment No.	Factor levels		
	A	B	C
1	1	1	1
2	1	2	2
3	1	3	3
4	2	1	2
5	2	2	3
6	2	3	1
7	3	1	3
8	3	2	1
9	3	3	2

The metal cutting process for the combination of 3 factors selected at 3 levels each were simulated using computer and the mean cutting force in Newtons (N) measured for each experiment. As the objective of the study is to find optimum combination of factor levels so that the cutting force generated is minimum, Taguchi's objective function smaller-the-better S/N ratio was selected. The ratio used is,

$$\ddot{u} = -10 \log\left[\frac{1}{N}\sum y_i^2\right]$$

The cutting force measured and S/N ratio (R_i) calculated using the above objective funcion is given in the following table.

Experiment No.	A	B	C	F_c	R_i
1	1	1	1	12.63	–22.02
2	1	2	2	18.01	–25.1
3	1	3	3	26.11	–28.3
4	2	2	2	12.50	–21.9
5	2	3	3	20.91	–26.4
6	2	1	1	37.99	–31.6
7	3	3	3	13.28	–22.4
8	3	1	1	28.77	–29.2
9	3	2	2	36.66	–31.3

For instance the S/N ratio R_i is arrived by squaring the value of F_c and taking the log of the quantity and then multiplying it by –10., i.e.,

$$R_1 = -10\log (12.63)^2 = -22.02.$$

Similarly for other eight experiments (R_i) values were obtained.

The average S/N ratio for factor levels R_{A1} is,

$$R_{A1} = \frac{1}{3}\left[R_1 + R_2 + R_3\right]$$

$$= \frac{1}{3}\left[-220 - 25.1 - 28.3\right] = -25.1$$

$$R_{A2} = \frac{1}{3}\left[R_4 + R_5 + R_6\right]$$

$$= \frac{1}{3}\left[-21.9 - 26.4 - 31.6\right] = -26.6$$

$$R_{A3} = \frac{1}{3}\left[R_7 + R_8 + R_9\right]$$

$$= \frac{1}{3}\left[-22.4 - 29.2 - 31.3\right] = -27.6$$

Similarly the calculated average S/N ratio values for other factor levels are,

$$R_{B1} = -22.1 \qquad R_{C1} = -27.6$$

$$R_{B2} = -26.9 \qquad R_{C1} = -26.1$$

$$R_{B3} = -30.6 \qquad R_{C1} = -25.7$$

Based on the above calculations, the following table indicating the factor levels and the average S/N ratio (R) was developed.

Factor	Levels		
	1	2	3
A	-25.1	-26.6	-27.6
B	-22.1	-26.9	-30.6
C	-27.6	-26.1	-25.7

The primary aim is to determine the best or the optimal combination of factor levels so as to ensure minimum of cutting force. The optimum level for each factor is the level that gives the highest value for R_i in the experimental region. Since –log is monotonely decreasing function, it implies that we should maximize R_i. From the values obtained for R_i, value –22.1 is preferable to –26.9 because –22.1 is greater than –26.9. Therefore the best setting for cutting speed (A) is A1, the best feed rate B1 and the best edge radius is C3. From this, the researcher concluded that the best setting of factor levels to achieve minimum cutting force is A1, B1, C3, i.e., cutting speed of 100 m/min. feed rate of 0.1 m/revolution and edge radius of 200 μm.

The researcher also conducted a full factorial experiment by simulation involving all the 27 combination of factor levels and the cutting force generated was measured in Newtons (N). Results of the simulated full factorial experiment for the 27 combinaton of factor levels is given below:

	C1			C2			C3		
	B1	B2	B3	B1	B2	B3	B1	B2	B3
A1	12.63	21.17	34.09	10.31	10.01	39.49	8.89	15.17	26.11
A2	13.42	24.02	37.99	12.50	20.54	30.54	10.13	20.91	27.05
A3	18.07	28.77	42.77	14.29	22.49	36.66	13.28	20.80	32.27

By mere inspection of the above table, it can be found that the optional combination of factor levels is A1, B1, C3, which agrees with the results using Taguchi's orthogonal approach.

Fortunately this study involves only three factors at 3 levels each and the number of combination for a full factorial design is only 27. Suppose an experiment involves 6 factors at 3 levels each, then 729 (3^6) experiments have to be carried out using full factorial design. But by adopting Taguchi's orthogonal array method, the number of experiments will get reduced to only 19, i.e., little less than 2.5% of the total number of experiments.

9.4 MULTIVARIATE ANALYSIS

9.4.1 Classification of Multivariate Techniques

All statistical techniques which simultaneously analyse more than two variables on a sample of observations can be categorized as multivariate techniques. The basic objective underlying multivariate techniques is to represent a collection of massive data in a simplified way. In other words, multivariate techniques transform a mass of observations into a lesser number of composite scores in such a way that they provide as much information as possible contained in the raw data obtained from a research study.

There are two types of multivariate techniques: i) for data containing both dependent and independent variables, and ii) for data containing several variables without dependency relationship. In the first category, there are statistical techniques like multiple regression analysis, multiple discriminant analysis, multivariate analysis of variance and canonical analysis. In the second category, techniques like factor analysis, cluster analysis, multidimensional scaling (MDS) and latent structure analysis are included.

In this section we will cover only regression analysis, correlation and curve fitting which are more useful for engineers and scientists.

9.4.2 Curve fitting

On many occasions, a researcher has to deal with two or more variables, which are functionally related and one would like to establish a relationship. Sometimes it is straightforward and easily discovered, but more often the functional relationship is complex and completely unknown. The first step in deriving an equation to approximate the desired relationship is to collect the data showing the corresponding values for the variables under consideration. Suppose in a heat treatment process, we consider the yield strength of the steel as a function of the temperature, we may let y as the yield strength and x equal to the temperature, the independent variable.

If we take a sample of n data as (X_1, Y_1) (X_2, Y_2)... (X_n, Y_n) and plot it on a Cartesian two-dimensional coordinate system, we will get a scatter diagram as shown in Figures 9.1 and 9.2. From the scatter diagram, it is often possible to visualize a smooth curve to approximate the data. It can be a straight line as shown in Figure 9.1 or a nn-linear curve as indicated in Figure 9.2.

The problem of finding equations for the approximating curves that best fit the given set of data is called curve fitting. The researcher has to select an appropriate curve to be fitted based on the scatter diagram. Some of the common types of approximating curves and their equation are given below.

i.	$y = a_o + a_1 x$	:	Straight line
ii.	$y = a_1 x_1 + a_2 x^2$	:	Parabola or quadratic curve

iii. $y = a_o + a_1x + a_2x^2 + a_3x^3$:   Cubic curve

iv. $y = a_o + a_1x + a_2x_2$:   nth degree curve

v. $y = ab^x$ or $\log y = a_o + a_1x$:   Exponential curve

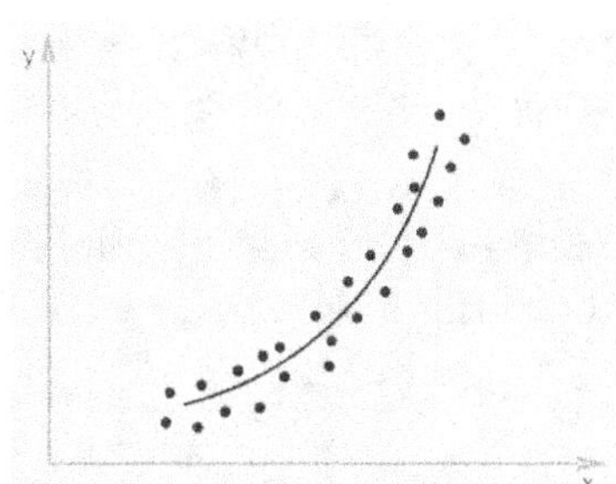

Figure 9.1   Linear relationship   **Figure 9.2**   Non-linear relationship

There are many such curves. The analyst can examine the scatter diagrams and compare the pattern with general shapes of curves given by different equations. Figure 9.3 shows the shapes of some of the curves.

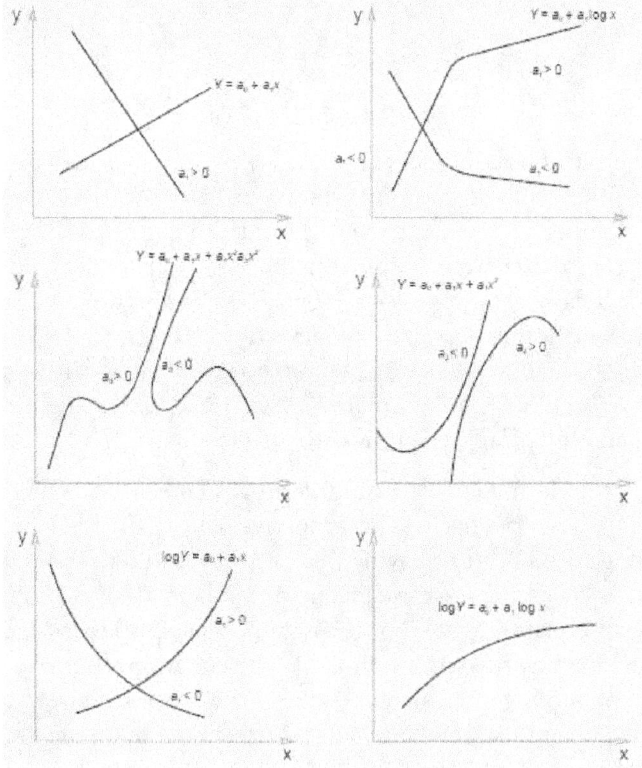

Figure 9.3   General shape of curves

Sometimes it may be useful to examine scatter diagrams of transformed variables using special graph paper, in which one or both scales are calibrated logarithmically. Special graph papers such as semi-log or log–log graph papers are available. For example, if the scatter diagram of log y versus x shows a linear relationship, we could use equation, $y = ab^x$ or $\log y = a_0 + a_1 x$ for log y vs x exponential curve. Likewise if one plots log y vs log x and if it appears linear, one can use a cubic logarithmic curve, $\log y = a_0 + a_1 \log x$.

To determine the best fit, one can decide intuitively or using a ruler or french curves and move it on the scatter diagram until it passes through the centre of the points. The method of fitting a curve by eye judgement is frequently used. In this approach an attempt is made to minimize the deviaions of the points from the prospective line.

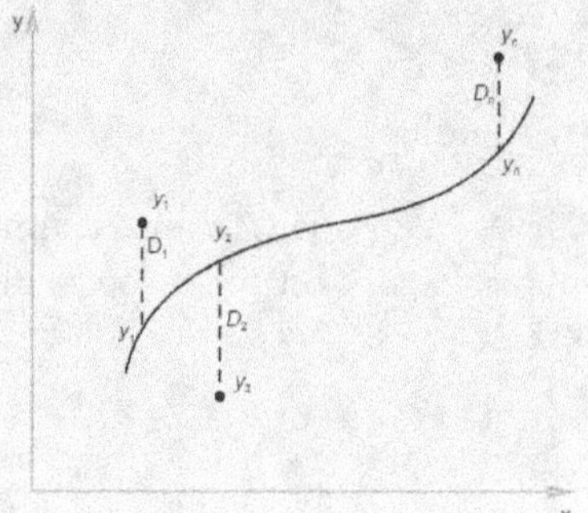

Figure 9.4   Fitting of curve

Consider the set of data points (x_1, y_1), (x_2, y_2), ... (x_n, y_n) and the approximating free hand curve fitted to the data points shown in Figure 9.4. For a given value of x_1, there will be difference between the value y_1 and the corresponding estimated value $\hat{y}_1$ determined from the curve. Let this difference $\left(y_1 - \hat{y}_1\right)$ be denoted as D_1 and it is referred to as a deviation, error or residual. It may be positive, negative or zero. Similarly, corresponding to the values x_2, x_3 ... x_n, deviations D_2, D_3 ... D_n can be obtained. A measure of the "goodness of fit" of the curve to the given data is provided by the quantity, $D_1^2 + D_2^2 + \cdots + D_n^2$, (i.e., sum of square of deviations). If this quantity is small the fit is taken as good and if it is large the fit is bad.

From the above information, one can infer that of all curves approximating a given set of data points, the curve having the property that, $D_1^2 + D_2^2 + \cdots + D_n^2$ is a minimum is called a best fitting curve. A curve having this property is said to fit the data in the least square sense and is called the least square curve. Mathematically this method is known as Method of Least Squares (MLS). For the data points (x_1, y_1), (x_2, y_2), ... (x_n, y_n), a least square line approximating the set of points has the equation $y = a_0 + a_1 x$, where a_0 and a_1 are constant representing intercept and slope of the line respectively. For the given set of data, the value of a_0 and a_1, can be calculated using calculus approach as explained in the next section.

9.5   SIMPLE REGRESSION AND CORRELATION

9.5.1   Simple Regression Analysis

Often on the basis of sample data, we may estimate the value of variable y corresponding to a given value of variable x. In engineering, the analyst may try to study the effect of temperature on yield strength, effect of stress on number of cycles to failure or time to rupture, etc. This can be accomplished by estimating the value of the dependent variable y from the least square curve which fits the sample data. The resulting curve is called a "regression curve" of y on x, the independent variable, since y is estimated from x. Similarly we can have a regression curve of x on y treating x as dependent variable and y as independent variable. The regression line or curve of y on x is called trend curve and is often used for purposes of estimation, prediction or forecasting.

Many a time the relationship between the dependent and independent variables may not be linear. However, it is possible to linearize the relationship through a simple transformation of variables. While doing so, it is important to perform all calculations within the transformed space and then re-transform the results to the original engineering units for final presentation.

Linear regression is appropriate, when an approximate linear relationship is contemplated between two measurable characteristics. The general form of the curve to be fitted must be chosen by the analyst after examination of data and the scatter diagram. In the case of two measurable characteristics x and y, we can consider the linear form, $y = \alpha_0 + \alpha_1 x + \varepsilon$ where x is the independent variable, y the dependent variable, α_0 the true intercept of the regression equation, α_1 the true slope and ε the measurement or experimental error by which it differs from the ideal linear relationship. The coefficients α_0, α_1 represent population parameters.

Once an assumption of approximate relationship is made, the problem becomes one of estimating the parameters and of the regression equation from the sample data points (x_1, y_1), (x_2, y_2), ... (x_n, y_n). The sample equation is, $\hat{y} = a_0 + a_1 x$ in which $\hat{y}$ is the predicted value of dependent variable and a_0 and a_1 are the sample estimates of regression coefficients α_0 and α_1. Certain assumptions underlie in using the method of least squares and they include,

- The errors are normally distributed with mean of zero and constant variance σ_ε^2
- The errors are independent of each other.

The method of least squares establishes the line through the data in such a way that the sum of squares of the vertical deviation of observations from the line is smaller than fo any other line that could be drawn through the data.

$$y = \alpha_0 + \alpha_1 x + \varepsilon$$

$$\therefore \ \varepsilon_i = a_0 + a_1 x_i - y_i \ \text{and} \ \sum \varepsilon_i^2 = \sum \left(a_0 + a_1 x_i - y_i \right)^2$$

In order to get the values of parameter a_0 and a_1 that result in best value of $\sum \varepsilon_i^2$ take the first derivative and equte it to zero

i.e.,

$$\frac{\partial \varepsilon_i^2}{\partial a_0} = 2 \sum \left(a_0 + a_i x_i - y_i \right) = 0$$

$$\frac{\partial \varepsilon_i^2}{\partial a_1} = 2 \sum \left(a_0 + a_i x_i - y_i \right) x_i = 0$$

By separating and rearanging we get,

$$\sum y_i = a_0 n + a_1 \sum x_i$$

$$\sum x_i\, y_i = a_0 \sum x_i + a_1 \sum x_i^2$$

Solving these two equations we get,

$$a_1 = \frac{n \sum xy - \sum x \sum y}{n \sum x^2 - \left(\sum x \right)^2} = \frac{\sum \left(x - \bar{x} \right)\left(y - \bar{y} \right)}{\sum \left(x - \bar{x} \right)^2}$$

$$a_0 = \frac{\sum y - a_1 \sum x}{n} = \bar{y} - a_1 \bar{x}$$

$$= \frac{\left(\sum y \right)\left(\sum x^2 \right) - \left(\sum x \right)\left(\sum xy \right)}{n \sum x^2 - \left(\sum x \right)^2}$$

The standard error of estimte of y for the data is calculated using the formulae

$$SEE = \left[\frac{\sum \left(y_i - \hat{y}_i \right)^2}{n - 2} \right]^{1/2}$$

$$= \left[\frac{\left(y_i - a_0 x_i - a_1 \right)^2}{n - 2} \right]^{1/2}$$

9.5.2 Simple Correlation Analysis

One has to understand that fitting of a straight line (or any other curve) to the data does not mean that the physical data or relationship are best described by that equation.

Mathematical operations always establish the value of parameters which give the best least squares fit to the equation form chosen. In order to assess whether the data are a "good fit" to the line or equation, we need the concept of "correlation". Correlation tells us how close the data points cluster around the curve or line. While regression defines a proposed relationship between the variables, correlation tells us how good that relationship is. Te various types of correlations are shown in Figure 9.5.

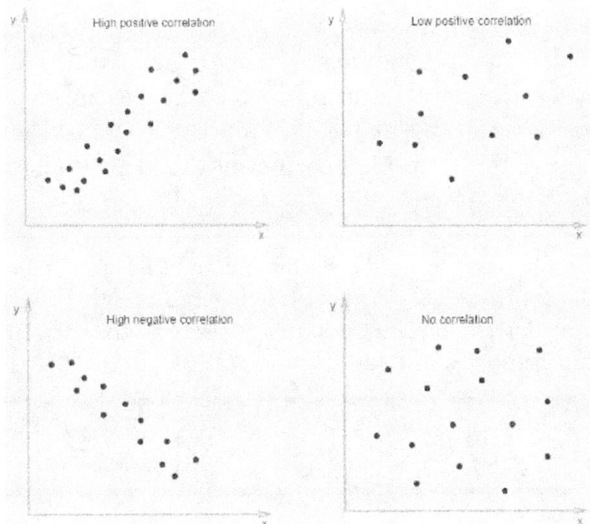

Figure 9.5   Types of correlation

A high correlation between variables simply indicates that they change their values in a related manner. But this does not prove or imply a cause-and-effect relationship based on the phenomena and variables involved. Regression analysis assumes that there is a cause and effect relationship between the dependent and independent variables, correlation makes no such assumption. The important point is that a researcher makes an assumption or hypothesis about cause-and-effect relationship between the variables on the basis of prior knowledge of the phenomenon studied when he does a regression analysis. This assumption may or may not be true. Correlation tells him how closely his data points cluster around the curve he has hypothesized. It does not tell him whether his basic assumption of cause and effect is valid. However, correlation analysis does tell him whether the data are consistent with his hypothesis.

Correlation coefficient is a number which ranges from –1 to +1. A coefficient of –1 means perfect negative correlation between variables. A coefficient of zero means no correlation and a coefficient of +1 means perfect positive correlation. Very seldom can one get any of these three values. Often only a fraction of that is obtained and it is necessary to conduct test for statistical significance. For a simple reression problem, the correlation coefficient R is given by

$$R = \frac{n\sum xy - \left(\sum x\right)\left(\sum y\right)}{\left[n\sum x^2 - \left(\sum x\right)^2\right]\left[n\sum y^2 - \left(\sum y\right)^2\right]^{\frac{1}{2}}}$$

Using the correlation coefficient one can also infer how well the linear equation arrived at fit the raw data. If R = ±1 then a perfect linear fit is inferred between x and y. In general, the closer the value of R to 1, the better is the linear relationship.

The total variation of y is defined as , i.e., the sum of the squares of the deviations of y from the mean . This variation comprises of an unexplained variation and an explained variation . The ratio of amount of the variation which can be explained by the regression equation, to the total variation observed is called the "coefficient of determination", which is given by the square of the correlation coefficient (R^2). This is illustrated by means of an example.

Example   The table below gives the results obtained from an experiment for the film thickness in Angstrom as a function of time in minutes that the metal specimen was at temperature. Fit a suitable regression line,also find the correlation coefficient and comment on the fit.

Time in min.	20	30	40	60	70	90	100	120	150	180
Thickness	3.5	7.4	7.1	15.6	11.1	14.9	23.9	27.1	22.1	32.9

This problem involves only two variables and hence a simple linear regression equationcan be fitted. The calculation table ca be set as indicated:

x	y	x^2	y^2	xy
20	3.5	400	12.25	70.0
30	7.4	900	54.76	222.0
40	7.1	1600	50.41	284.0
60	15.6	3600	243.36	936.0
70	11.1	4900	123.21	777.0
90	14.9	8100	222.01	1341.0
100	23.9	10000	571.21	2390.0
120	27.1	14400	734.41	3252.0
150	22.1	22500	488.41	3315.0
180	23.9	32400	571.21	4302.0
Σ 860	156.6	98800	3071.2	16889.0

The number of observatios (n) = 10,

$$\Sigma x = 860 \qquad \therefore \bar{x} = \frac{860}{10} = 86$$

$$\Sigma y = 156.6 \qquad \therefore \bar{y} = \frac{156.6}{10} = 15.66$$

The slope a_1 is calclated using the relation

$$a_1 = \frac{n\sum xy - \left(\sum x\right)\left(\sum y\right)}{n\sum x^2 - \left(\sum x\right)^2}$$

$$= \frac{(10)(16889) - (860)(156.6)}{(10)(98800) - (860)^2} = \frac{34214}{248400}$$

$$\therefore a_1 = 0.138$$

The intercept a_0 is calculated as

$$a_0 = \bar{y} - a_1 \bar{x}$$

$$= 15.66 - 0.138(86) = 3.79$$

The regression equation fitted is,

$$y = 3.79 + 0.138\,x$$

Correlation coefficient (R) is calculated using the formula,

$$R = \frac{n\sum xy - \left(\sum x\right)\left(\sum y\right)}{\left\{\left[n\sum x^2 - \left(\sum x\right)^2\right]\left[n\sum y^2 - \left(\sum y\right)^2\right]\right\}^{1/2}}$$

$$= \frac{(10)(16889) - (860)(156.6)}{\left\{\left[(10)(98800) - (860)^2\right]\left[(10)(3071.2) - (156.6)^2\right]\right\}^{1/2}}$$

$$= \frac{34214}{\sqrt{(248400)(6189)}}$$

$$= 0.873$$

The coefficient of determination R^2 is,

$$R^2 = (0.873)^2 = 0.762$$

This implies that 76.2% of variation in film thickness (y) with changes in time (x) can be explained by the equation $y = 3.79 + 0.138x$. We can also state that the linear equation is reasonably the best fit for the experimental data to express relationship between film thickness and time since $R = 0.878$.

9.6   MULTIPLE REGRESSION AND CORRELATION

9.6.1   Multiple Linear Regression

When the dependent variable of interest is influenced by more than one independent variable and if one is interested in establishing relation between the dependent variable and the independent variables, the statistical tool of multiple regression is often useful. The most common and mathematically tractable form of this method requires that the model be assumed be linear in the coefficients that will be adjusted to minimize the sum of squares of all the residuals. Mathematically the relation must be expressible in the form as in the case of two independent variables.

$$y = a_0 + a_1 x_1 + a_2 x_2 + \varepsilon$$

This model is linear because it is linear in the coefficients a_0, a_1, a_2 and ε represents the random error. The above model describes a regression plane in the three-dimensional space of y, x_1 and x_2. The parameter or coefficient is te intercept of this plane as shown in Figure 9.6 for the model $y = a_0 + a_1 x_1 + a_2 x_2 + \varepsilon$. The coefficients a_1, a_2 are referred to as partial regression coefficients. The coefficient a_1 represents the change in the mean response corresponding to a unit change in x_1 when x_2 is held constant. Similarly, a_2 represents the change in the response corresponding to a unit change in x_2 when x_1 is held constant. The model shown in Figure 9.5 is a first order multiple regression model because the maximum power of the variable in the model is one. Also shown is an observed data point y_1 and the error for that point.

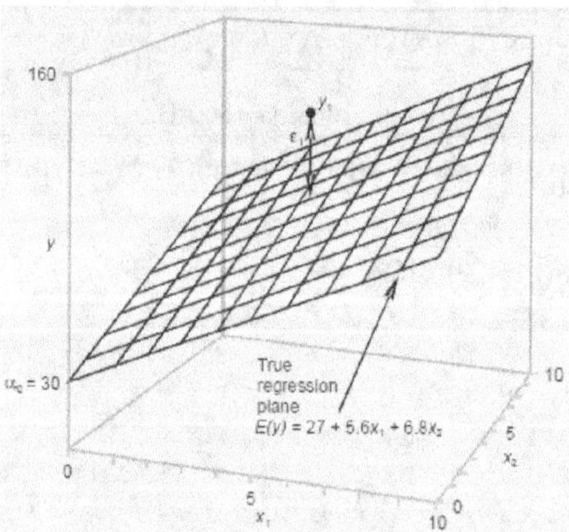

Figure 9.6   A first-order multiple regression model

The best values of the coefficients a_0, a_1 and a_2 will again be found by using a least square criterion. First, we should decide as best as we can on the form of the model and then gather data consisting of observation of y and x_j. We then define the residual at each observation equal to the difference between the measured y and the model predicted y. We add the squares of all the residuals forming the total error (E). To minimize E, we choose the a_j so as to make all the partial derivatives of E with respect to each a_j equal to zero. Suppose, the mult-regression equation involve only two independent variables,

$$y = a_0 + a_1 x_1 + a_2 x_2 + \varepsilon$$

$$E = \sum (y_j - a_0 - a_1 x_{1j} - a_2 x_{2j})^2$$

$$\frac{\partial E}{\partial a_0} = -2\sum \left(y_j - a_0 - a_1 x_{1j} - a_2 x_{2j}\right) = 0$$

If there are m coefficients to be found, we need at least m observations to be able to find a_j. Usually we have more observations than coefficients, therefore, we need to use least squares methods of linear algebra to solve these equations and get the values of a_j. After getting the values for a_j and a_0 the desired multi-regression equation can be obtained. If more than two independent variables are involved manual computation is very difficult and tedious. Hence, one has to resort to the use of software packages.

Computer software packages are available to perform the multivariable calculations indicated above. "Microsoft Excel" is one such package. A computer program for multiple regression and correlation analysis will typically include ANOVA of the regression. This will consist of multiple regressions, total residual sum of squares, degrees of freedom and as well as residual mean squares. The total sum of squares is an expression of the total amount of variability among the y_j values the regression sum of squares expresses the variability among the y_j values attributable to the regression being fit and the residual sum of squares tells us about the variability of y_j still emaining after fitting the regression The ANOVA table will be,

Source of variation	Sum of (SS) squares	d.f.	Mean square (MS)
Regression	$\sum \left(\hat{y}_j - \bar{y}_j\right)^2$	m	$\dfrac{\text{regression SS}}{\text{regression df}}$
Residual	$\sum \left(y_j - \hat{y}_j\right)^2$	$(n - m - 1)$	$\dfrac{\text{regression SS}}{\text{regression df}}$
Total	$\sum \left(y_j - \bar{y}_j\right)^2$	$(n - 1)$	

The total number of data prints is n (i.e., total number of y values and m is the number of independent variables of the regression model).

If we assume y to be functionally dependent on each of the x's, then we are dealing with multiple regression. If no such dependence is implied, then any one of the $(M = m + 1)$ variables could be designated as y for the purpose of utilizing computer program, this being the case of "multiple correlation".

The ratio $R^2 = \dfrac{\text{regression SS}}{\text{total SS}}$ or equivalently $R^2 = 1 - \dfrac{\text{regression SS}}{\text{total SS}}$ is the "coefficient of determination" for multiple regression or correlation or the "coefficient of multiple determination". In the regression situation, it is an expression of the total variability in y attributable to the dependence of y on all the x_j's as defined by the regression model fitted to the data. As in the case of correlation, R^2 may be considered to be the amount of variability in any one of the m variables that is accounted for by correlating it with the other $(m-1)$ variables. The square root of the coefficient of determination is referred to as "multiple correlation coefficient", i.e., $R = \sqrt{R^2}$

9.6.2 Multiple Non-Linear Regression Models

In many engineering problems, we may encounter with non-linearity. For instance, a second-order model with two independent variables is given by:

$$y = a_0 + a_1 x_1 + a_2 x_2 + a_3 x_1^2 + a_4 x_1 x_2 + a_5 x_2^2$$

Such models are needed to reflect non-linearity and second-order effects. This model is a second-order model because the power of the terms in the model is two. A cross product term $x_1 x_2$ is also included in the model. This term represents an interaction effect between the independent variables x_1 and x_2. Interaction means that the effect produced by a change in one independent variable (x_1) on the response depends on the level of the other independent variable (x_2). The regression surface in a three-dimensional coordinate system is shown in the Figure 9.7. These regression models are used in Response Surface Methodology (SM) to find the optimum value of the response. Consider another example

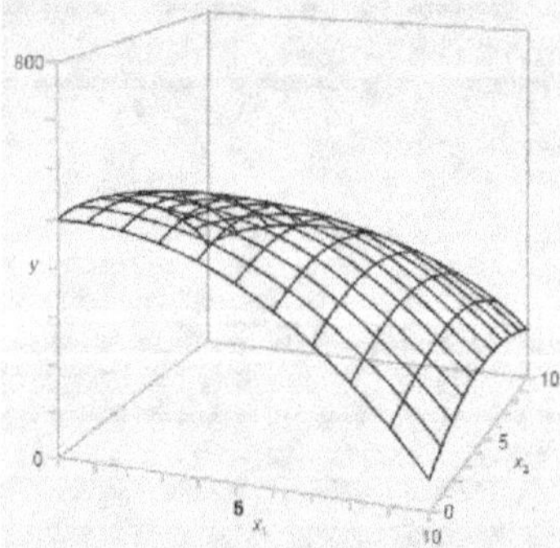

Figure 9.7 Regression surface

In this case, the independent variables are x_1, x_2, x_3 and x_4. To make this fit into linear multiple regression model, one has to do some transformation by defining new variables as, $x_a = x_1$, x_2 and $x_b = x_3^3$. Similarly in the case of another example,

$$y = a_1 m_1^{x_1} \cdot m_2^{x_2} \cdots m_n^{x_n}$$

the equation could be modified by taking logarithm on both sides

$$\log y = \log a_1 + x_1 \log m_1 + x_2 \log m_2 + \ldots + x_n \log m_n$$

The use of multi-regression analysis is explained through a case study included at the end of this section.

Standardized computer programs are usually used to determine the parameters of this type of equations. When the analyst use the "canned" statistical program in computer, due care has to be exercised. The researcher should understand the capability of the statistical package and what it really does. The user should take time and trouble to read the backup documentation for the program thoroughly, or if necessary, consult experts, and then use it.

Case Study

This is actually an extension of case study narrated in Chapter 5, wherein the researcher has developed a multi-regression equation based upon dimensional analysis linking tracking resistance with sx other variables for the polymeric insulation material newly developed.

Symbol	Description	Units	Variable
CTI	Comparative index	volt	x_1
DES	Dielectric strength	Kv/mm	x_2
SR	Surface resistivity	ohm	x_3
AR	Arc resistance	sec	x_5
VR	Volume resistivity	ohm-m	x_5
TR	Tracking resisitance	min	y

In order to counter-check the multi-regression equation developed using dimensional analysis, the researcher conducted physical experiments under laboratory conditions by varying the independent variables *CTI, DES, SR, AR* and *VR* and measuring the dependent variable, TR. The details regarding independent and dependent variables used are given in the table already given.

Thus, the multi-regression equation considered can be expressed as a function,

$$y = f(x_1, x_2, x_3, x_4, x_5)$$
$$= D(x_1)^a (x_2)^b (x_3) c (x_4) d (x_5)^e$$

Experimental measurement for both dependent and independent variables is given in the table below:

TR	CTI	DES	SR ($\times 10^{14}$)	AR	VR ($\times 10^{13}$)
116	475	27.56	2.60	363	4.70
127	495	30.6	2.65	370	4.10
135	501	33.5	2.68	376	3.58
146	505	35.7	2.73	379	3.18
218	530	35.9	3.43	395	3.20
163	509	33.3	2.78	3.82	3.10
235	536	24.1	3.46	400	3.28
175	512	32.5	2.82	386	3.05
193	515	32.0	2.86	390	2.99
205	517	31.5	2.90	394	2.95
213	518	31.4	2.93	397	2.90
222	521	31.0	2.95	401	2.87
228	518	30.5	2.97	408	2.85

As more than two independent variables are involved, with the help of software the multi-regression analysis was carried out. The exponents calculated are as follows.

$$a = 0.884, b = -0.674, c = 0.978$$

$$d = 0.843, e = -0.974, \text{ and the constant } D = 33442.$$

The regression equation arrived at is,

$$y = 33442\,(x_1)^{0.884}\,(x_2)^{-0.674}\,(x_3)^{-0.978}\,(x_4)^{0.843}\,(x_5)^{-0.974}$$

i.e.,

$$TR = 33442(CTI)^{0.884}(DES)^{-0.674}(SR)^{0.978}(AR)^{0.843}(VR)^{-0.974}$$

The fitted equation was checked for the error with predicted values of y and the range of errors varied between -10.2% to 5.2% which is reasonably good. This example illustrates the practical utility of multiple regression analysis in research studies. The predictions made using regression equation developed based on experimental data were checked with the predictions made by the regression equation developed using dimensional analysis and found that the agreement was good.

9.6.3 Multiple and Partial Correlations

The degree of relationship existing between three or more variables is called multiple correlations. The fundamental principles involved in multiple correlations are analogous to simple correlation discussed in the earlier section. For easy understanding, let us consider the case of three variables x_1, x_2 and x_3. Suppose we want to keep x_1 as dependent variable and x_2 and x_3 are independent variables, then the multiple linear regression equation could be written using special notations as,

$x_i = a_{1.23} + a_{12.3}x_2 + a_{12.2}x_3$ where $a_{1.23}$ is the intercept of the regression plane, and are coefficients. Because of the fact that x_1 varies partially due to variation in x_2 and partially due to x_3, we can refer $a_{12.3}$ and $a_{13.2}$ as partial regression coefficients of x_1 on x_2 keeping x_3 constant and x_1 on x_3 keeping x_2 constant respectively.

We can arrive at linear correlation coefficient between x_1 and x_2, x and x_3, x_2 and x_3 as explained under simple correlation using the relation

$$r_{12} = \frac{n\sum x_{1j}x_{2j} - \left(\sum x_{1j}\right)\left(\sum x_{2j}\right)}{\sqrt{\left(n\sum x_{1j}^2 - \left(\sum x_{1j}\right)^2\right)\left(n\sum x_{2j}^2 - \left(\sum x_{2j}\right)^2\right)}}$$

where, $j = 1, 2...n$ the sample data points. Similarly r_{13} and r_{23} can be calculated. The values of r_{12}, r_{13} and r_{23} may vary between -1 and $+1$. The coefficient of linear multiple correlation can be computed from the formula,

$$R_{1.23} = \sqrt{\frac{r_{12}^2 + r_{13}^2 - 2r_{12}r_{13}r_{22}}{1 - r_{23}^2}}$$

The coefficient of multiple correlations such as $R_{1.23}$ lies between 0 and 1. The closer it is to 1, the better is the linear relationship. If the coefficient of multiple correlations is 1 the correlation is perfect. The quantity $R_{1.23}^2$ is called the coefficient of multiple determination.

It may be necessary to measure the correlation between a dependent variable and one particular independent variable when all other variables involved are kept constant. This can be obtained by introducing a coefficient of partial correlation. Suppose we denote the coefficient of partial correlation between x_1 and x_2 keeping x_3 constant as $r_{12.3}$, then we can calculate,

$$r_{12.3} = \frac{r_{12} - r_{13}r_{23}}{\sqrt{\left(1 - r_{13}^2\right)\left(1 - r_{23}^2\right)}}$$

we can find interesting relationships connecting multiple correlation coefficients and various partial coefficients. For example the coefficient of multiple determinations is related to r_{12} and $r_{13.2}$ as,

$$R_{1.23}^2 = 1 - \left(1 - r_{12}^2\right)\left(1 - r_{13.2}^2\right)$$

Computer software are available and they provide outputs, viz., multiple correlation coefficient, multiple regression equation, partial correlation coefficients and ANOVA details once the sample data is input properly.

While using multiple regression techniques the analyst has to be careful about the problem of multi-collinearity. Multi-collinearity is on account of high degree of correlation between independent variables. In such cases the analyst can enlarge the sample size or consider dropping of one of the highly collinear variable and also take dequate care in selecting the independent variables to estimate the dependent variable.

10
Analysis and Interpretation of Data

10.1 INTRODUCTION

Once the collection of data for a study is completed, then the researcher has to draw inferences followed by report writing. To draw meaningful inferences, proper analysis of data gathered is necessary so that their interpretation is on the desired lines. If this is not done properly, it may lead to misleading conclusions and the whole purpose of doing research may get vitiated. It is only through interpretation that researchers can identify relations and processes that underlie their findings. Generally, tools available for analysis include statistical, graphical and numerical methods. Many of the techniques have already been covered in the previous chapters. A general approach to data analysis may include:

- Examine the data for consistency
- Perform necessary appropriate statistical and other analyses
- Estimate the uncertainties in the results
- Anticipate the results from theory
- Correlate the data gathered.

10.2 DATA CHECKING

10.2.1 Treatment of Outliers

During data collection phase and as a part of data analysis, one should continue to be on the watch out for invalid data, using whatever data checking procedures that are appropriate to the study. Weeding out "bad" data needs to be a continuing effort at all stages of the study. Undetected spurious data, once they get into a sophisticated data analysis procedure, can bias results and interpretation in many ways which may be unclear. Tools for rejection of spurious data include "treatment of outliers" and use of "running graphs".

A researcher must perform preliminary screening of data before carrying out the desired analysis. The purpose of such screening is to detect and discord any data points which are questionable, so that they do not improperly bias the analysis. These decisions must be made prudently because the wild points usually called as "outliers" may actually be valid data and may be the cause for unexpected but potentially important behaviour.

The outlier may crop up due to various factors such as misreading of instrument, voltage fluctuations, poor connection, etc. However, if one takes more readings, it becomes easier to detect outlier data that might occur. If we could estimate the standard deviation from sample mean values, then any value beyond ±3s from average would reasonably be a candidate for rejection. Though this is only a guideline, one cannot escape from the responsibility of exercising our best engineering judgment. If readings are rejected, we should make note of this and explain the basis of our judgment in the report.

It is quite likely that a different context of our outlier decisions is found when we plot a graph using data of independent and dependent variables as shown in Figure 10.1. Points 'a' and 'b' are both outliers; however the point 'a' is more easily judged as wild than point 'b'. If a wild point is located at the ends of a curve, though it is an accurate one, it may indicate the beginning of a change in the behaviour, and it cannot be reckoned as error. Unless we have clear evidence of mistakes, such points should be tested by extending the range of the experiment so that these questionable points become interior rather than exterior points.

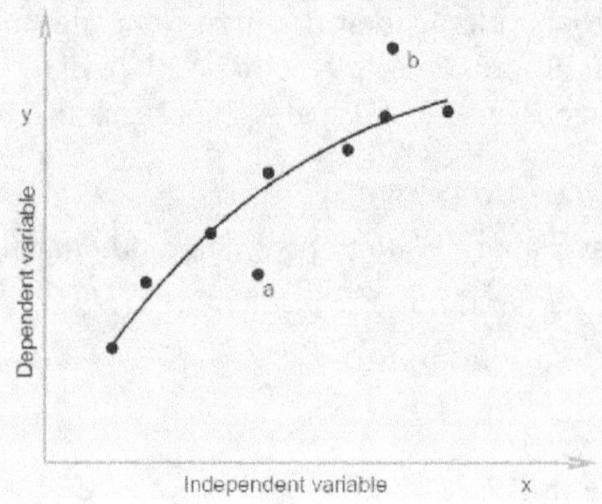

Figure 10.1   Outliers

10.2.2   Plotting of Running Graphs

Running graphs provide a reliable method of detection of "odd behaviour". As we gather data points, we should plot appropriate graphs. Compared with a table, a graph is a much more sensitive and reliable detector. By keeping these running graphs updated as we add each reading, we can often detect problems and malfunctions quickly. These graphs are usually not a part of our final data sheet and will usually make on a rough notebook. Here, we need to plot the actual data wherever possible, rather than processed data which may turn out to be the computed results of our study.

For example if we are recording water level in a stand pipe and the discharge rate of an orifice at the bottom of the stand pipe for the purpose of producing a graph of flow rate versus pressure drop for the orifice, our running graph could use raw water level data in centimetres rather than take the time to compute the pressure in Pascal from the water level.

10.2.3 Some Useful Suggestions

The data checking methods actually used will depend on the nature of experiment or study. However the following general suggestions will be useful to researchers.

It is better to repeat part of all those runs which appear questionable though it is expensive and time-consuming. If this can be justified, it will be most convincing if any random effects are allowed to be different from the original run when we replicate it. The decision as to whether to undertake a replication and how comprehensive it should be will depend on the extent of doubt we have about the original run and how important the questioned results are to the success of the study.

When we graphically display data relating to two variables, the extrapolation of the curve to the origin may be a useful checking method. For example when we plot efficiencies of machines against load on the machine, it should extrapolate to zero at zero load. Similarly when we study properties of mixture of components, these properties should extrapolate to the properties of a component when the percentage of other components goes to zero. We should always look for and use any such known relations from the knowledge of the phenomena studied to check the behaviour of data.

One can also employ physical laws and principles to check whether data is consistent with itself. Conservation of energy tells us that if we have measured all the inputs and outputs of energy to a process operating at a steady state, then the sum of all the inputs must equal the sum of all the outputs. If our calculation of these inputs and outputs show a discrepancy, we may have some faulty data.

The principle of conservation of mass can be used in a similar manner, if we are measuring material inflow and outflow rates. In practice, such balance checks may not be possible or totally convincing if there are significant unmeasured losses such as leaks in the system. In general, we should consider carefully all the available theoretical information about our study and make use of those aspects that allow us to partially or totally check the validity of our data. This should be done before we undertake any large-scale overall data analysis associated with the objective of the study/research made.

10.3 DATA ANALYSIS

10.3.1 Statistical Analysis

Parameter estimation and comparison   When we expect some randomness in our data, the analysis method involves statistical considerations. If our study is one of establishing the numerical value of some measurable quantity, we must choose the percentage of the mean value we wish our confidence interval to be and for the confidence level we want. We may also need an estimate of the coefficient of variation. Especially when data is yet to be collected, the estimate of coefficient of variation must be based on past experience or one should resort to outright guess work. Using the statistical tools, we can also determine how large a sample is needed to maintain the desired level of accuracy.

In the case of raw data, the analysis consists merely of determining the sizes of confidence intervals for the chosen confidence levels. For example, if we have a sample size of 10, mean value of 6.8 and standard deviation of 0.47, the 95% confidence interval is 6.8, ± 0.33. Suppose our best estimate of the measured value is 6.9, we are 95% sure that it is lies between 6.47 and 7.13, i.e, within the confidence interval.

One may try to decide whether one material or process is better than an alternative one by comparing average value or standard deviation. Comparison of the average values is made by using confidence intervals, by using the difference between two mean values or by using the t-test. For comparing two standard deviations, F-ratio test could be used to reach decisions as to whether the two values are about the same or significantly different. The various statistical tools have been covered in the previous chapters 6,7 and 8.

The significance of difference between two process means can be analysed using either Z-test or the t-test. But, when we want to analyse more than two means, the ANOVA techniques enable us to conduct simultaneous tests and it is an important analysis tool.

Use of regression analysis In respect of problems where we have to find a relation between a single independent variable and a dependent variable, we can use the simple regression analysis technique. Preliminary data analysis should include plotting of the scatter diagram and then selecting an appropriate equation to fit the data. If we have a physical theory which predicts a definite mathematical relation between the two quantities, we can check for the consistency of graph with the theory. If visual evaluation indicates that the data could follow the predicted straight line, parabola, sine wave, etc. then we can use the regression analysis software to find the specific curve that best fits the data.

Suppose a theoretical prediction is available for the type of curve to be expected, but the data shows a large scatter, then the interpretation becomes less clear-cut. In spite of larger scatter, if we can still visually notice a trend in the data, which does not contradict the theory, then of course we should get a best fit regression using the theoretically predicted type of curve. Thereafter we can graph the residuals for this fit to see if they are largely random, or else we can detect a trend. If no trend is noticed, we may have found the best model. If a trend is noticed its nature may give us some clues about how to adjust the form of the model to get a better fit. A final choice of the model is based on criteria such as visual goodness of fit, numerical value of R^2, and numerical values of individual residuals relative to the practical importance of different regions of the curve.

In order to study the relation between one dependent and two or more independent variables, we can resort to data analysis using multiple regression techniques. The logical starting point for developing a regression model is to identify a suitable functional relation based on the knowledge of the phenomenon studied. Regression methods can be used to find the best-fit-surface in the case of two independent variables adopting the usual criteria for evaluating the quality of fit. As usual, data with large noise makes the choice of the best model more difficult.

If no theoretical prediction is available to guide us, three-dimensional plots of raw data are possible but are almost useless for the purpose of choosing a model form, if there are many points and/or there is much scatter.

Multiple regression models not only enable us to study the main effect of independent variables but also the interaction effect. For example consider the regression equation,

$$Y = a_o + a_1 x_1 + a_2 x_2 + a_3 x_3 + a_4 x_1 x_2 + a_5 x_2 x_3 + a_6 x_1 x_3 + a_7 x_1 x_2 x_3$$

The main effect of variable x_1 would be the model term $a_1 x_1$ while the term $a_4 x_1 x_2$ would be called an interaction between $x_1 x_2$. That is, interaction refers to a situation where the effect on the dependent variable of changes in one independent variable depends upon the value of some other independent variable.

When we study models with large number of independent variables, concepts of factorial or fractional factorial experiment and orthogonal arrays become useful. They are not only important in experiment planning but also helpful in data analysis.

Use of dimensional analysis   In Chapter 5, we studied how the technique of dimensional analysis could be used to develop relation between the factors involved in a phenomenon when the exact relationship between them are unknown or not well-established. By using dimensional analysis the various factors responsible for the phenomenon can be grouped into a few and each group as a whole can be varied instead of varying each individual factor to study the phenomenon. This reduces the experimental time and effort required for the analysis.

The technique of dimensional analysis has extensive application in analysing problems related to fluid flow in hydraulic engineering, thermal and heat transfer in mechanical engineering, aerodynamics in aeronautical and automobile engineering, etc. Also this technique can be applied to compare the relationship established based on physical experiments. To illustrate this, a case study based on a research work has been explained in Chapter 5. Thus this is a useful technique for the researchers to do data analysis.

Use of response surface methodology   Response surface methodology (RSM) is a collection of mathematical and statistical techniques used for modelling and analysis of problems in which the response is influenced by several variables coupled with the objective of optimizing this response. When we have to deal with more than one response variable, we have no other alternative, but to find a compromise optimum that does not optimize only one response. Certain problems have constraints also and so the experimental design has to take into account these constraints. Thus the goal of RSM is not only to identify optimum but also the factor levels that yielded the optimum result.

Experimental study plays an important role in science and engineering and several types of experiments are conducted in real-world problems. When treatments are from a continuous range of values then the true relationship between the dependent response y and the independent variable x may not be known. Especially if more than two independent variables are involved, the resulting relationship is not linear and it may be a surface. The relationship between independent variables and the response can be expressed in the form of a function. An approximation of the response function, $y = f(x_1, x_2 \ldots x_n) + \varepsilon$, is called response surface mehodology (RSM), where $(x_1, x_2 \ldots x_n)$ are independent variables, y is the response and is the error term. There are three RSMs, viz., first order, second order and three-level fractional factorial.

In the case of two independent variables x_1 and x_2, with y as the response, the response surface can be expessed by the relation $y = f(x_1, x_2) + \varepsilon$. The error term represents measurement error if any and also the unaccounted variations in the response. It is a statistical error and it is assumed to be normally distributed with mean zero and variance s^2. The response surface for the two variables case is shown in Figure 10.2.

Figure 10.2a represents the response surface and Figure 10.2b shows the contour plots which are isocontour lines of x_1 and x_2 pairs that have the same response value y. While graphs are useful tools to understand the response surface, it is difficult to visualize graphically when we deal with more than two variables.

The major problem is one of finding the design points which gives the optimum for the response, i.e., the value of x_1 and x_2 which gives the maximum or minimum value depending upon the objective of the problem. Once we develop a response surface graphically, it is easy to locate the optimum point by examining the figure. But, however in reality, the problem may not be so simple due to complexity of the phenomena there may be more number of variables. Hence, we have to resort to use o statistical techniques.

Response surface methodology has extensive applications in research studies. It is a powerful statistical procedure which employs factorial analysis, regression analysis and search techniques, etc. to determine the optimum point on the response surface and also to identify the factor (variable) levels that have contributed to the optimum. Another important use of RSM is that it helps to guide the researcher to move from the starting experiment to the next experiment and so on to finally reach the optimum point.

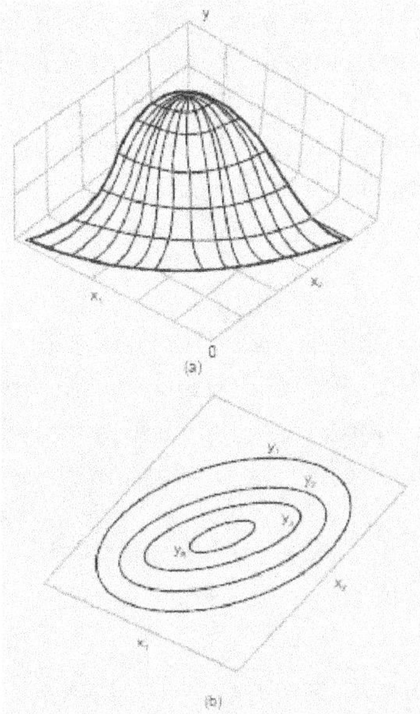

Figure 10.2 Response surface

Analysis should be meaningful Any analysis should bring forth the meaning contained in the data. The meaning will have its own detailed interpretation for each specific case. For instance the study may involve estimation of average value of a parameter or the variability of the parameter that can be used to study a system's performance. The performance parameter tells us whether the overall design goals have been met or not. Average values are used to calculate the nominal value of the parameter, while the standard deviation is used to estimate variability around the average. Comparison of mean or average values can be used to evaluate alternative designs, process, materials, etc.

Normally simple linear regression and multiple regression models are used to establish relations between process outputs such as quality, cost, yield and the process inputs which affect these outputs. Once such relations have been identified, we can use them to effect adjustments to the process aimed at improving operations. The operating points not only maximize some desirable output but also make it less sensitive to changes in input values. This input can be used to guide relaxation of specifications on component parts or materials, so as to reduce costs, still maintaining quality. The meaning of data which defines a process model thus lies in our better understanding of how the process works.

10.3.2 Graphical Data Analysis

Nowadays interactive software for personal computers renders regression methods easy, yet graphs play an useful role for simpler problems. Most of these methods are intended for exploring the relations between two variables by using a simple x-y graph. To speed up the work, special graph paper with various kinds of distorted scales (log – log, linear – log, Gaussian probability, etc.) is often used. When such paper is not available, one can get the equivalent effect by using linear scale paper for plotting not the raw data, but rather the desired function (such as logarithm) on the graph. In fact this is what the most modern computerized versions of these methods do, using regression analysis to define the best line.

Let us consider the relation of the form $y = cx^a$. Taking logarithm on both sides log $y = \log c + a \log x$ which is equivalent to $y = b + mx$. We can graphically implement this transformation by plotting Y (log y) and X (log x) on the linear graph paper or y and x on log – log paper. By visual inspection we can fit a best straight line to our data and then judge whether the fit is acceptable. If the fit is acceptable, we get numerical values for the fitting constants c and a by graphical measurements. If we have used a computer to fit the best line, its equation would automatically be decided by the regression procedure. If the fit is not acceptable, we need to analyse faults to get some clues as to the next type of relation to try.

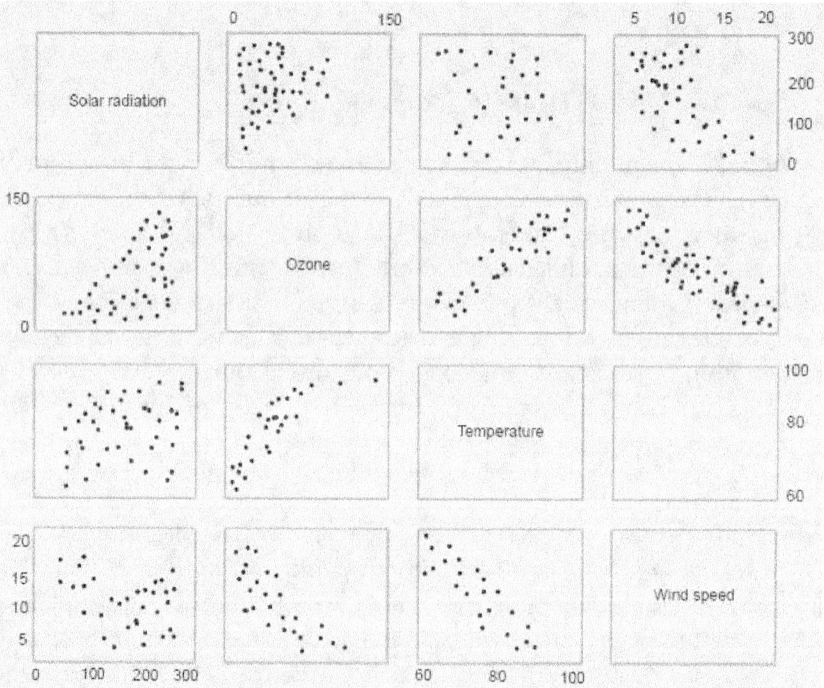

Figure 10.3   Scatter plot matrix

Sometimes a best fitting curve for the transformed values may not be best one for actual values in the least square fitting process, so we have to look at the fit quality for the actual values before deciding whether the fit is acceptable. This can be done by introducing Gaussian random noise to the y values and compare the noisy data (YG + NOISE) and the fit to the noisy data (YGFIT) using actual x and y values rather than the logarithm used in the fitting process. By doing this we can check whether the use of logarithms has created any problem in our case.

When one dependent variable with multi-independent variables are involved, computerized methods using regression analysis are generally preferred. However, a sequence of graphs of the dependent variable versus each of the independent variables will of course reveal which independent variable is dominating and what the shape of this dependency looks like. Scatter plot matrix for multivariable graphical data analysis is generally used for this type of analysis.

The scatter plot matrix can be applied to models with any number of variables. If the total number of variables (independent and dependent) is m, then one can make ordinary x, y graph between any two of them, there being a total of $m(m - 1)/2$ such graphs. Though there is no novelty in making such graphs, the contribution of the scatter plot matrix concept lies in how the graphs are organized and presented. The example shown in Figure 10.3 explains how the actual air pollution data has been depicted to explain this raphical data analysis method.

The level of ozone concentration, wind speed, solar radiation and air temperature were measured for a period. These data are plotted in a scatter plot matrix form. In this, the graphs are arranged in rows and columns such that each row or column has all the graphs relating a certain variable to all the others. In the instant case, the upper row has graphs of solar radiation plotted against each of the other three variables. The third column from the left gives plot of solar radiation, ozone and wind speed against temperature. All the other rows and columns have to be similarly interpreted. Combining some qualitative knowledge of pollution chemistry with information from the various graphs enables one to reach some useful conclusions about this pollution process. Some of the commercial statistical software packages do provide this facility.

10.3.3   Numerical Data Analysis

Most data analysis methods employ a combination of statistical, graphical and mathematical (numerical) tools. While software can be written to perform any calculations one may need, the scope of such methods is essentially unlimited. The time frame and effort may be the constraints. But, nowadays, a large number of readymade dedicated software packages are available to help the researchers and analysts. They are user-friendly and menu-driven.

The commercially available statistical software packages can easily perform many statistical analyses such as computation of means, standard deviations, correlation coefficients, t-test, analysis of variance, analysis of covariance, multiple regression, factor analysis and various non-parametric analyses. These are just a few of the programs and sub-programs that are available in many packages.

A researcher can use conventional text-based languages such as 'C' to write the application programme, or higher-level more user-friendly languages such as LabVIEW.

The LabVIEW uses graphic software modules called virtual instruments (VIs) that have front-panel user interface and block diagram programs. By connecting icons for these modular VIs in a block diagram, one can easily create, modify, combine and exchange VIs to form more sophisticated VIs. A compiler is used to generate the machine code so that VIs execute on a computer at speeds comparable to compiled 'C' programs.

The routines available broadly includes the following.

- Fourier and spectral analysis
- Digital filters and windows
- Wave-form analysis
- Wave-form generation
- Statistical analysis
- Vector and matrix algebra
- Numerical analysis

In every new edition new features are being incorporated. Depending upon the need, the analyst can choose and analyse the data gathered.

MATLAB is another important computational software extensively used by researchers for analysis. MATLAB is available for a large number of computer systems and has great degree of platform independence. The MATLAB environment provides the user with an interactive workspace in which to enter commands and view results. With MATLAB one can

i.   define variables, enter their numerical values and perform calculation interactively

ii. create vectors and matrices and perform operations on them using the many built-in MATLAB functions.

iii. visualize the calculated results by plotting them on the screen and printing a hard copy.

iv. use the built-in MATLAB programming language to create a calculation procedure consisting of a number of related steps.

v.   use the text editor to write and save a script file, which contains a specific series of MATLAB statements that one wants to use in the future.

vi. use the script files to define functions that are of use in future work.

vii. employ the tool boxes of functions developed by others and make available as script files.

Online help is available within the MATLAB environment. A menu provides a table of contents as well as an index. Every new version has many new features incorporated.

10.4   INTERPRETATION OF RESULTS

10.4.1 Interpretation in Research

Interpretation is regarded as the task of drawing inferences from the data collected after an analytical, experimental or simulation study. In fact it is a quest to get the broader meaning of research findings. The task of interpretation has two purposes, viz.,

    i.   To establish continuity in research by linking the results of a given study with those of others.

    ii.   To establish some explanatory concepts from findings.

In one sense, interpretation is concerned with relationships within the data gathered, which partially overlap analysis. Also, it extends beyond the data of the study to include the results of other research, theory and hypothesis to establish correlation.

The interpretation is always related to the objective and scope of the study envisaged. It helps the researcher to gain better comprehension and understanding of factors that explain the observations of the researcher in the study. Also, it provides a theoretical conception, which can serve as a guide for the future researchers.

10.4.2 Need for Interpretation

Interpretation of results is essential for the simple reason that the usefulness and utility of research findings lie in it. Moreover, it is considered as a basic component of the research process due to the following reasons.

    i.   Only through proper interpretation can a researcher well-establish the abstract (theoretical) principles that lie beneath their findings. Through this, they can link up their findings with those of other studies having the same concept or theory, and thereby can predict concretely. Subsequent inquiries can test these findings later on. In this way the continuity in research can be maintained.

    ii.   Interpretation leads to the development of explanatory concepts and methods that can serve as a guide for future research studies. It opens up intellectual adventure and stimulates the quest for more knowledge.

    iii.   The researchers can appreciate better only through interpretation of the causative factors influencing their findings which enable others to understand the real significance of their research work.

    iv.   In some exploratory type of research studies, the interpretation of findings often results into hypothesis which requires further experimental research study. Thus, the interpretation enables transition from exploratory to experimental research. Since an explanatory study does not have a hypothesis initially the interpretation of findings of such a study leads to the formation of hypothesis.

10.4.3   Correlation with Scientific Facts

A research work may be analytical, experimental, simulation or some combination of these. The theoretician strives to explain or predict the results of experiments on the basis of analytical models which are in accordance with fundamental physical principles that have been well-established over the years. Suppose the outcome of an experiment does not fit into the scheme of existing physical theories, the researcher may start doubting his/her experimental data and then their appropriate theories. In some cases, the earlier findings are modified to take into account the results of the new experimental data, after ensuring the validity of the data. In any event all physical theories must eventually rely upon experiment for verification. Hence, this necessitates consistency of results to be achieved.

The researcher must give reasonable explanation of the relations found and must interpret the relationship in terms of underlying scientific processes or principles. The researchers should try to find out the common factors if any in their diversified research findings. Any findings based on theoretical, or experimental study should be explained with the science involved. In fact, this is the technique of how generalization should be done and concepts be formulated.

Any extraneous information, collected during the study, must be taken into account while interpreting the final results of the research study, extraneous aspects sometimes prove to be key factors in understanding the problem under consideration.

A statistical analysis is appropriate only when measurements are repeated several times. If this is the case, we have to make estimates of parameters such as average, standard deviation, etc. In such cases where the uncertainty of data is to be prescribed by statistical analysis, calculations should be performed using appropriate distribution. Justification for using a particular distribution should form part. Also, this can be used to determine level of confidence using appropriate level of significance. The number of measurements made should be justified using confidence levels and confidence interval.

Before using regression techniques for analysis and interpretation the analysts have to carefully review the theory appropriate to the subject. Besides he/she should gather necessary scientific information that will indicate the trends that the results may take. Important dimensionless groups, pertinent functional relations and other information may lead to fruitful interpretation of data. Interpretation of interaction effect between factors indicated by the regression model should be backed by the theoretical knowledge about the behaviour of interacting factors pertaining to the phenomena under study.

Modelling, whether mathematical or simulation, is an art and science. The researcher develops a model of an existing system or its part or a hypothetical system taking into account the objective and scope of the study. The modeller may ignore some of the actual features of the system under study and abstract from the real situation only those aspects that are relevant to the study objectives. Thus, models are essentially simplifications and abstractions of real situations. In that process the researcher makes many assumptions to simplify his/her model.

The rationale behind the various assumptions in developing the model has to be explained based on the results/test of small experiments, experience of other researchers quoted in their published work, expert's views and logical reasoning by the researcher. Each assumption has to be justified as how it does not affect the model's result for the purpose of the study undertaken.

Sometimes we assume linear relationship between two variables even though we may suspect or even know that the true relationship is curvilinear. We assume that at least over a limited range of variation such an approximation will be satisfactory. For, instance, an electrical engineer works with models of circuits in which the values of resistors, capacitors, etc. are assumed to have constant values. In reality electrical characteristics of many of these components vary as a function of temperature, humidity, age, etc. but yet simplified assumptions are made for the purpose of modelling. The mechanical engineer works with models in which gases are perfect, pressures are adiabatic, and conduction is uniform. For majority of practical cases, such approximations are good enough and yield usable results. All that the researchers have to do is to justify their explanation based on conceptual and other scientific knowledge they command.

Before attempting to draw conclusions based on the output results of experiments or simulation models, the researchers should establish the validity by comparing with real-world results and expert opinion. If there are certain limitations, they have to be explained and shown as to how they have not affected the findings and conclusions. The experience of other researchers may also be quoted to support their justifications.

While analysing results and making interpretations, consideration has to be given to aspects such as accuracy vs precision, and error vs uncertainty. While accuracy refers to the agreement between a measurement and true or correct value, precision refers to the repeatability of a measurement. Error refers to the disagreement between true and accepted value whereas uncertainty of a measured value is an interval around that value such that any repetition of the measurement will produce a new result that lies within this interval.

Depending upon the nature of the study, experiment and objective, etc. a researcher has to carry out analysis applying the aspects mentioned in the previous paragraphs. But, to establish error, one must know the correct value. But one does not know the true value ahead of time, especially when the central objective is to discover new things. An experienced analyst may assume that the experiment is not in error as this is the only choice available. Only future studies can corroborate this. Therefore instead of assuming they are correct they should explain as to how at each step of their study, they had exercised checks with available scientific background of knowledge.

10.5   GUIDELINES IN INTERPRETATIONS

A researcher should keep in mind that even if the data are properly collected and analysed, wrong interpretation would lead to inaccurate conclusion. Therefore, it is necessary that the task of interpretation be performed with patience and in an objective manner. The following guidelines may be useful to the researcher while making interpretation.

i.   In the first instance researchers must invariably make sure that a) the data are appropriate, trustworthy and adequate for drawing inferences b) the data is homogeneous and, c) proper analysis has been made through statistical methods.

ii.   One should be cautious about the errors that may creep in while interpreting results. Errors arise due to false generalization and/or due to wrong interpretation of statistical measures such as extension of findings beyond the range of observations, identification of correlations with causation, etc.

iii.   Another pitfall is the tendency to affirm that definite relationships exist on the basis of confirmation of a hypothesis. It only indicates that there is no valid statistical reason to reject the hypothesis at the chosen level of significance. Moreover it does not confirm the validity of the hypothesis. The researchers should remain vigilant about all such things so that false generalizations may not take place. They should be well-equipped with and must know the correct use of statistical measures for drawing inferences concerning their study.

iv.   One should be aware that both interpretation and analysis go together and cannot be distinctly separated. Hence, they should feel the task of interpretation as a special aspect of analysis and accordingly must take all precautions concerning reliability of data, computational checks, validation and comparison of results, etc.

v.   It is necessary that researchers be conscious that their task is not only to make sensitive observations of relevant occurrences, but also to identify and discord the factors that are initially hidden to the eye. This will enable them to do their job of interpretation on proper lines. As the coverage may be restricted to a particular time, a particular context and conditions, they should avoid broad generalizations and the results must be framed within their limits.

vi.   Researchers should ensure constant interaction between initial hypothesis or premise, empirical observations and theoretical conceptions. Only such interaction between theoretical orientation and empirical observations provide opportunities for originality and creativity.

vii.   One has to be very careful while using packaged simulation software. The in-built logic may not be readily available and often it is used as a black box. Perhaps the manuals provide only information regarding input of data and extraction of results. Therefore, a researcher has to be careful while making assertions based on his/her findings from such study and comparing it with others. The software used by another person may be a different one or a different version.

viii. Most often physical experiments are carried out in laboratory environment, where it may not be possible to replicate the real environment. This has to be kept in mind while interpreting the results based on experiments.

ix. While using bought out instruments for use in experiments, precaution has to be exercised with regard to the sensitiveness, accuracy of the instrument. Alternatively due consideration has to be made in interpretation base on the readings made using those instruments.

11
Accuracy, Precision and Error Analysis

11.1 INTRODUCTION

Researchers use several instruments in their experimental studies to suit the experimental set-up developed for the specific purposes and the procedure devised by them. All these depend on the nature of study. To ensure good results of their study, the researchers have to be concerned with aspects such as uncertainties, accuracy, errors, etc., relevant to their study. While the instruments should be accurate and precise, the experimental set-up and procedure adopted should ensure less variability and bias. Notwithstanding the various precautionary measures adopted by the analyst, some errors are bound to occur on account of many uncontrollable factors.

The occurrence of error cannot be avoided often and hence error analysis becomes an important aspect. Researchers can perform an error analysis even before the conduct of the experiment by carrying out sensitive analysis of variables or factors to identify which are more sensitive to the outcome of the experiment. This will enable them to take extra precautions with regard to selection of variables and their levels. After conducting the studies, the researchers can perform error analysis to arrive at the error associated with their results for reporting. This will improve the credibility of the research studies made. Before proceeding further, it will be useful to the analyst if a proper understanding of the terminology related to above aspects is made.

11.1.1 Accuracy

Accuracy is concerned with the agreement between measurement and true value. For instance if a clock strikes twelve when the sun is exactly overhead, the clock is said to be accurate. Here the indication of the clock as twelve and the phenomenon it is meant to measure (location of sun at zenith) are in agreement. With regard to instruments, accuracy can be defined as the ability of the instrument to record the true value of a measured variable under the reference conditions. It is measured as a percentage of full scale reading, percentage of true value or percentage of scale span. For instance a pressure gauge of 100 kPa having an accuracy of 1% would record within ±1kPa over the entire range of scale.

Often in experimental studies both accuracy and bias are used synonymously. The accuracy of a set of measurements quantifies the difference in those test results from an accepted reference or standard value. To enable us to make quantitative statements about absolute accuracy or bias, we need to know the reference points. But, unfortunately such reference points are seldom available in the real-world situation. Suppose we are confident that an instrument is within certain plus or minus range of the true value we say in such cases the plus or minus range expresses the uncertainty of the instrument readings.

11.1.2 Precision

Precision indicates the repeatability of measurement and it does not require us to know the correct or true value. Suppose on each day a clock reads exactly 11.30 AM for several years when the sun is at zenith, we say this clock is very precise. The true meaning of noon may not be important here because, we only care whether the clock is giving a repeatable result. In the case of measuring devices, it is defined as the degree of exactness for which a device is designed or intended to perform. Precision is composed of two characteristics, viz., conformity and significant figures. Conformity refers to the conforming of quality of an instrument to the standards prescribed by the national authorities of a country. The significant figures of a measured quantity are the digits which are meaningful in the quantity measured. Several conventions are being followed as how to indicate the numbers as well their significant digits.

The terms precision and accuracy are often used interchangeably in measurements using instruments. But actually there is a difference between these two terms. Let us consider the measurement of a known temperature of 100°C with a thermometer. The six readings taken are 103°, 104°, 102°, 103°, 102° and 104°. While the average value is 103°, the maximum deviation from the average reading is ±1° in 100°. Thus, the scale of the instrument can be calibrated to read ±1°C. But with reference to accuracy, the readings are not accurate. In the above case, the accuracy of the instrument is only (104 – 100)/100, i.e., 4%. Thus, we reckon the precision as ±1% and the accuracy as 4%.

A precise measurement may not necessarily be accurate and vice versa. We can illustrate this statement with an analogy from shooting practice on a target as shown below.

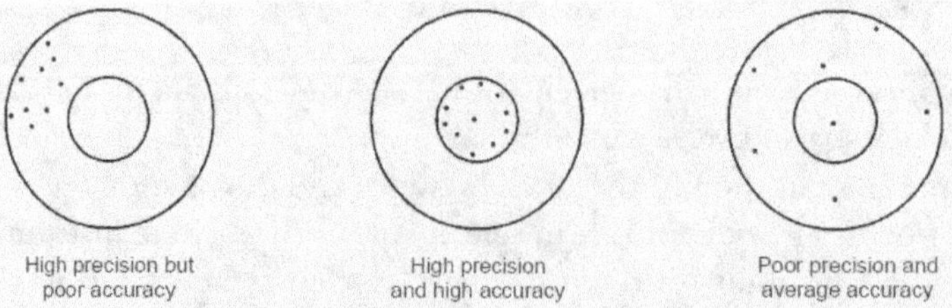

In the context of an experimental study, precision can be defined as the reciprocal of the sample standard deviation (1/s). Since the standard deviation is a measure of the scatter, a decrease in the scatter of a test result will end in a smaller standard deviation which ultimately leads to an increase in precision. A statement concerning the precision of a test method provides a measure of the variability that can be expected between test results when an appropriate method is adopted.

11.1.3 Uncertainty and Variability

The terms uncertainty and variability are used sometimes interchangeably. However, there are subtle differences between them. A typical dictionary meaning of the word "uncertainty" is "the condition of being in doubt". Uncertainty can be considered as the limits within which the "true" engineering result such as the average life of a component lies. Uncertainty is commonly described in terms of confidence limits or tolerance limits or prediction limits.

Variability is an important element of uncertainty calculation. The meaning of the term can be inferred from its root word 'variable'. As the behaviour of many variables or factors are random in nature, they can be described using metrics such as average or standard deviation or sample variance. Another commonly used measure is coefficient of variation.

11.2 REPEATABILITY AND REPRODUCIBILITY

A research study is considered as a contribution to the existing literature and it is supposed to enlarge the existing knowledge base in the chosen field of research study. It is quite likely that other researchers and interested persons may refer the study already made and try to repeat the experiment. Therefore, any experimental work should be designed in such a way that they are repeatable by another researcher from any part of the world and should be able to reproduce the results following the same procedure and conditions. The results may not agree fully, but it should be within the confidence limits specified by the original researcher. The aim of the other researcher may be to counter-check the procedure followed by the previous researcher so that he/she can adopt the same for their study or modify and build further on it for some other purpose. Thus repeatability and reproducibility assume greater significance in any research study.

Repeatability is concerned with the closeness of agreement among the number of consecutive measurements of the output for the same value of input under the same operating conditions. Repeatability may also be referred to as test–retest reliability in the case of repetition of tests in experiments. The variation in measurements may be due to human errors or due to certain defects in the instruments used. If the variation between measurements is smaller than some agreed limit, we can take it that the measurements are repeatable. Certain conditions are to be satisfied when we consider repeatability and they include the following.

- the procedure followed for measurement must be the same.
- the observer should not be changed, i.e., the same observer has to be used.
- the measuring instrument used must be the same one and it should be used under the same conditions.
- the repetition should be over a short period of time.

The repeatability of an experiment can be expressed by means of a coefficient known as "repeatability coefficient". This coefficient is a precision measure, which satisfy a value below which the absolute difference between two repeated test results can lie with a probability of 95 per cent. In fact, the standard deviation under repeatability condition is a part of precision and accuracy. It may be specified in terms of units for a given period of time.

Reproducibility may be defined as the closeness of agreement among the repeated measurements of the output for the same value of input under the same operating conditions over a period of time. Thus, the main difference between repeatability and reproducibility is one of time duration among other conditions. Perfect reproducibility implies that there is no drift in the output over a long period of time. Reproducibility is one of the main principles of the scientific method. It ensures that the test or experiment can be reproduced or replicated accurately by some other researcher working independently. In fact, it represents the ability or robustness of the experiment to be repeated. The values obtained by the researcher who repeated the experiment of somebody should be similar to the original experiment.

However, there are differences between reproducibility and repeatability. While repeatability measures the success rate in successive experiments conducted by the same experimenter, reproducibility relates to the agreement of test results despite the fact that they are carried out by different experimenters, test apparatuses, laboratories and locations. Normally, it is reported as standard deviation (s). Both reproducibility and repeatability are considered as a measure of the closeness with which a given input may be measured again and again.

This concept can be extended to the research study. Assuming that a researcher planned his/her study properly, and designed the experiment or model with utmost care and dexterity thereby making it robust, then reproducibility and repeatability of results under identical conditions by another researcher, should not pose any problems. This is an essential feature of a research work. John Clearbout of Stanford University proposed the idea of reproducible research. He is of the view that the ultimate product of research is the paper along with the full computational environment used to produce the results in the paper such as code, data, etc. necessary for reproduction of the results and building upon the research.

11.3   ERROR DEFINITION AND CLASSIFICATION

11.3.1 What is an Error

We can define error as the disagreement between a measurement or a true or accepted value. Our knowledge about the physical world is obtained by conducting experiments and making measurements. While doing this it is imperative to appreciate that all measurements of physical quantities are subject to uncertainties. Though it is not possible to measure anything exactly, effort should be made to reduce the error as much as possible, but it is always there. In order to draw valid conclusions the error must be indicated and dealt with properly. If the result of a measurement is to have meaning, it cannot consist of measured value alone. Therefore, an indication has to be made to show how accurate the result is.

Though the terms "error" and "uncertainty" are used synonymously, there is semantic difference between these two terms. Holman is of the view that it is better to speak of experimental uncertainty, instead of experimental error, because the magnitude of error is always uncertain. The task before the experimenter is how to specify the uncertainty in analytical form. A reasonable approach to define experimental uncertainty is by specifying the possible value that the error may have.

As already mentioned, an error refers to the disagreement between a measurement and the true or accepted value, but, in reality the true value is not known with certainty. Therefore, when taking measurements, occurrence of error, which is related to uncertainty is unavoidable. The result of any physical measurement has two essential components: i) a numerical value giving the best estimate possible of the quantity measured, and ii) the degree of uncertainty associated with this estimated value. For example, a measurement of the diameter of a steel rod would yield a result such as 40 ± 0.1 mm.

If measurements are taken repeatedly, the values obtained will differ and a particular value of the result cannot be preferred over the rest. Although it is not possible for us to do anything about such error, we can characterize it. For instance, the repeated measurements may cluster tightly together or they may spread widely and this pattern can be analysed systematically.

11.3.2 Types of Error

Systemic error or bias error   One can broadly classify errors as systemic error and random errors. Systemic errors are errors which may tend to shift all measurements in a systematic way so that their mean value is displaced. This may be attributed to things such as incorrect calibration of equipment, improper use of equipment consistently or failure to properly account for a certain effect on the part of the researcher. While large systemic errors can be eliminated in a good study, small systemic errors will always be present. In scientific studies, we need several independent confirmations of experimental results because the equipment behaviour may be affected at different places thereby introducing systemic errors. The systemic errors are called as "bias errors" or "fixed errors".

We can estimate fixed error or bias error by means of theoretical calculations. Suppose we want to measure the temperature of hot gas flowing through a pipe by means of a mercury thermometer. The bulb portion which is at the bottom of the thermometer is exposed to hot gases while the upper tube portion is exposed to the surroundings. Therefore, the temperature we measure using this instrument does not reflect the real temperature of the gas and an error is bound to be there despite the fact that we take many readings. However, the fixed error in this case can be estimated knowing the thermal properties of the gas and the glass thermometer.

To explain the concept of systemic or bias error we can consider a simple experiment. Suppose we perform an experiment with a simple pendulum to estimate the value of the local gravitational acceleration constant g. The relevant equation linking the period of oscillation (T) and other parameters such as the length (L) of the pendulum keeping the initial position as the position when the pendulum is at rest

$$T = 2\pi \sqrt{\frac{L}{g}} \tag{1}$$

This equation can be expressed in terms of $\hat{g}$, representing the esimate of g as,

$$\hat{g} = \frac{4\pi^2 L}{T^2} \tag{2}$$

The usual procedure we follow in conducting the experiment is to measure the length (L) of the pendulum and make repeated measurements of period (T), starting each time from the same initial position. We measure values of T repeatedly and then average it. The measured values for L and average T are substituted in equation (2) to obtain an estimate of g, i.e., $\hat{g}$.

In this experiment there are three possible sources of systemic or bias errors. It is possible that the experimenter may measure the pendulum length wrongly from the point where the string is attached rather than from the centre of the mass of the bob. Also there is a possibility of using an improper device to measure L. This is not a random error; it occurs each time the length is measured. Systemic error could also occur if the experimenter has consistently not counted properly the back and forth motion of the pendulum inadvertently to obtain an integer number of cycles. There can be an inherent defect in the stopwatch used to measure the time, and it may either read too large or too small a value. It is quite likely that the experimenter has not kept the bob at the exact initial position every time when the experiment is repeated.

In fact, the experimenter can study the effects of systemic error possibilities before conducting the experiment by doing a sensitivity analysis. This can be done by introducing small perturbations in L and T and studying theoretically using equation (2) the effect of each one on $\hat{g}$. This approach will enable the experimenter to identify the relative effect of each parameter on so that he/she can devise better and accurate methods for measuring these parameters. If perturbation (small disturbances) of a factor has negligible effect, perhaps the effect can be ignored and the procedure/instrument that was originally contemplated to be used for measurement can be adopted.

Random errors   Random errors refer to the fluctuation in measurement from one measurement to the next. It has been found generally that the results they yield are distributed about some mean value and there are variety of reasons for the occurrence of such errors. They are,

- Lack of sensitivity in equipment
- Presence of noise, i.e., extraneous disturbances
- Imprecise definition
- Random electronic fluctuation in instruments
- Various influences of friction

The random error measurements are displaced in an arbitrary fashion whereas systemic errors displace measurements in a single direction. Random errors are unavoidable and must be lived with.

The previous example of pendulum experiment can also be used to explain the random error or precision error that will occur. While the experimenter repeatedly measures the period of oscillation of the pendulum, he/she may obtain different values for each measurement. These fluctuations are random and small in magnitude. The fluctuations may be due to the reaction time of the experimenter in operating the stopwatch, differences in observing when the pendulum has reached its maximum angular travel, etc. All these random causes may interact and produce the variations or fluctuations in the measured quantity T.

It should be noted that the fluctuations mentioned above are not due to systemic or bias errors and moreover these variations are unpredictable. But, however these random fluctuations tend to follow some pattern. Based on studies it has been found that they tend to follow normal or Gaussian distribution. It is usual to assume that these random fluctuations follow a normal distribution and are amenable for statistical analysis.

Though the case example cited is a simple one, the concepts could very well be applied to any major study wherein the final response is obtained through an equation and the experimental results have to be substituted into the equation to get the derived response. In order to minimize the systemic error, it is necessary to study the effect of each input parameter on the outcome. This can be checked even before the conduct of experiment so that a researcher can properly plan the design of the experiment, instruments to be used and experimental set-up. The random error can be estimated through statistical analysis and reported. However, by adopting proper sampling method, sample size, etc., the random error can be minimized.

11.4 ANALYSIS OF ERRORS

11.4.1 Impact of Error Propagation

As indicated in the previous section in many cases it is likely that the result of an experiment will not be measured directly; rather it will be calculated based on several measured physical quantities. If the error occurs in each measurement, it may have cascading effect in the overall results of the experiment. This necessitates that an estimation of the cascading or propagation effect of errors has to be made.

If x_1 and x_2 are two measure quantities with errors and respectively, the combined error (δz) in the worst possible case will be obtain10'ed by direct calculaton as indicated below.

$$\delta z = (x_1 + x_2) = (x_1 + \delta z_1) + (x_2 + \delta z_2) = (x_1 + x_2) + (\delta z_1 + \delta z_2)$$

$$\delta z = (x_1 - x_2) = (x_1 + \delta x_1) - (x_2 + \delta x_2) = (x_1 - x_2) - (\delta x_1 + \delta x_2)$$

For instance, we may want to calculate power from the relation $P = (V)$ (I) where V is the voltage and I is the current. Suppose the voltmeter has an error of ± 2 V and the ammeter ± 0.2 A and if the measurement shows 100 V and 10 A, the nominal value of the power is $100 \times 10 = 1000$ watts. If we take the worst possible variations in voltage and current, we could the calculate the power,

$$P_{max} = (100 + 2)(10 + 0.2) = 1040.4 \text{ Watts}$$

$$P_{min} = (100 - 2)(10\ 0\ 0.2) = 960.4 \text{ Watss}$$

This method of calculation shows the error or uncertainty in the power is $+4.04$ per cent and -3.96 per cent. It is quite unlikely that extreme situations arise both in voltage and current measurements. However, this will provide some guidance with regard to inspection of raw data. If the error calculated is beyond the extreme limits, one has to recheck the data more thoroughly. In case the variables are independent it is likely that the error in one variable will cancel out some of the error in the other. Therefore the average of the error in Z will be less than the sum of the errors in its parts.

11.4.2 Error Propagation Estimation

We know that the randomness inherent in measuring a quantity follows a normal or Gaussian distribution. This implies that if a quantity is measured 100 times, 95 (i.e., confidence limit of 95%) of times the quantity will cluster within the limit of twice the standard deviation (i.e., 2σ). Suppose, a measured quantity x differs from its exact quantity

x' and the uncertainty involved is δx using the concept stated above, we can mathematically express it as $(x - x') = 2\delta x$. At 95% confidence level, 5% of the measurement will be inconsistent with the concept which leads to a discrepancy of 5%.

Assume that we measure two values x_1 and x_2 using different devices and the magnitude of the absolute error involved are δx_1 and δx_2 the relative error will be indicated by $\delta x_1 / x_1$ and $\delta x_2 / x_2$ respectively. Error propagation can be estimated as follows:

i. The square of the total uncertainty in the sum or difference is the sum of the squares of individual absolute errors.

If the relation between z and x_1, x_2 is $z = x_1 + x_2$ then

$$(\delta z)^2 = (\delta z_1)^2 + (\delta x_2)^2$$

If the relation between z and x_1, x_2 is $z = x_1 - x_2$ then

$$(\delta z)^2 = (\delta x_1)^2 + (\delta x_2)^2$$

Thus, if the measured quantity of x_1 is 6.5 with uncertainty ± 0.4 and x_2 is 5.5 with uncertainty of ± 0.3, then the total measured quantity of x_1 and x_2 will be $(6.5 \pm 5.5) \pm \sqrt{(0.4)^2 + (0.3)^2}$.

ii. The relative uncertainty in a product or a quotient will be the square root of the sum of the squares of the relative errors in the individual factors.

If the relation between z and x_1, x_2 is a product, i.e., $z = (x_1)(x_2)$ then

$$\left(\frac{\partial x}{x}\right)^2 = \left(\frac{\partial x_1}{x}\right)^2 + \left(\frac{\partial x_2}{x}\right)^2$$

If the relation between z and x_1, x_2 is a quotient, i.e., $z = x_1/x_2$ then

$$\left(\frac{\partial x}{x}\right)^2 = \left(\frac{\partial x_1}{x}\right)^2 + \left(\frac{\partial x_2}{x}\right)^2$$

iii. The relative error on an uncertain quantity raised to an exponent is the exponent times the relative error, i.e., $z = (x)^n$

$$\left(\frac{\partial z}{z}\right) = n\left(\frac{\partial x}{x}\right)$$

The above-stated aspects are subjected to certain assumptions and limitations, viz.,

i. errors are small compared to the measurements

ii. errors are statistically independent, and

iii. underlying distribution which these errors represent are normally distributed.

11.4.3 Using Partial Derivatives

Suppose we want to calculate the result of an experiment from a set of measurements and the final result is expressed as proposed by Kline and McClintock.

$$z = z(x_1, x_2...x_n)$$

The error or uncertainty in the calculated result can be expressed with the same probability as were used while estimating the errors in the measurement. We can use the root sum square combination of the effects of each of the individual inputs. The overall error or uncertainty will be,

$$\delta z = \left[\sum_{i=1}^{n} \left(\frac{\partial z}{\partial x_i} (\delta x_i) \right)^2 \right]^{\frac{1}{2}}$$

where, x_i, 1, 2 ...... n is a set of n measurements, δx_i is the error or uncertainty in x_i, δz is the overall uncertainty.

Each term in the above equation represents the contribution made by the error in one variable (δx_i) the overall error in the result (δz). The partial derivative of z with respect to x_i is the sensitivity coefficient for the result z with respect to the measurement x_i. The above equation holds good so long as each measurement is independent and follows normal distribution and in addition the uncertainties have the same probability (i.e., odds). If there are only two independent measurements x_1 and x_2, then

$$\delta z = \left[\left(\frac{\partial z}{\partial x_1} \delta x_1 \right)^2 + \left(\frac{\partial z}{\partial x_2} \delta x_2 \right)^2 \right]^{\frac{1}{2}}$$

In case we want to express the uncertainty estimate as a fraction of reading, rather than in engineering units, then the relative uncertainty or error can be expressed as,

$$\frac{\delta z}{z} = \left[\left(\frac{\partial x_1}{x_1} \right)^2 + \left(\frac{\partial x_2}{x_2} \right)^2 \right]^{\frac{1}{2}}$$

For example in the case of the pendulum experiment referred earlier, the relative error of g $\left(\text{i.e., } \dfrac{\delta \hat{g}}{\hat{g}} \right)$ can be arrived at as given below. $\delta(\hat{g})$ is the overall error in g and is the estimate of g.

$$\text{Estimate of } \hat{g} = \frac{4\pi^2 L}{T^2}$$

If $\delta(\hat{g})$ is the overall error (total error) then,

$$\delta(\hat{g}) = \left\{ \left[\frac{\delta \hat{g}}{\partial L}(\delta L) \right]^2 + \left[\frac{\partial \hat{g}}{\partial T}(\delta T) \right]^2 \right\}^{\frac{1}{2}}$$

where (δL) and (δT) represents error in L and T respectively. Substituting the corresponding values we get the relative error as,

$$\frac{\delta(\hat{g})}{\hat{g}} = \sqrt{\left(\frac{\delta L}{L} \right)^2 + \left(\frac{\delta T}{T} \right)^2}$$

The concepts stated above can be explained by means of a simple example. Suppose an analyst desires to measure the power dissipated across a resistor with the nominal stated value as 10 Ω ± 1%. The voltage measured and the current values are 100 V ± 1% and 10 A ± 1%. We can calculate the power dissipated using the relation $P = (V)(I)$.

Power dissipated $(P) = (V)(I)$

$$\frac{\partial P}{\partial V} = I \text{ and } \frac{\partial P}{\partial I} = V$$

If δV and δI are the error terms then

$$\frac{\delta P}{P} = \left[\left(\frac{\delta V}{V} \right)^2 + \left(\frac{\delta I}{I} \right)^2 \right]^{\frac{1}{2}}$$

Inserting the numerical values we get,

$$\frac{\delta P}{P} = \left[\left(\frac{1}{100} \right)^2 + \left(\frac{0.1}{10} \right)^2 \right]^{\frac{1}{2}}$$

$$= \left[(0.01)^2 + (0.01)^2 \right]^{\frac{1}{2}} = 1.414\%$$

We can use error estimation to guide us in the selection of an appropriate method of measurement in the experiment. For instance in the above problem if the error is calculated using te relation, power $(P) = V^2/R$ then,

$$\frac{\delta P}{\delta V} = \frac{2V}{R} \text{ and } \frac{\delta P}{\delta R} = \frac{V^2}{R^2}$$

$$\frac{\delta P}{\delta V} = \frac{2V}{R} \quad \text{and} \quad \frac{\delta P}{\delta R} = \frac{V^2}{R^2}$$

$$\therefore \frac{\delta P}{P} = \left[4\left(\frac{\delta V}{V}\right)^2 + \left(\frac{\delta R}{R}\right)^2 \right]^{\frac{1}{2}}$$

$$\left[4(0.01)^2 + (0.01)^2 \right]^{\frac{1}{2}} = 2.36\ \%$$

By comparing with the previous approach and this method, we can decide that the first method of power measurement using only voltmeter and ammeter is better than measuring resistance.

11.4.4 By Perturbing Variables

In many cases reduction of data is a complicated affair and calculation of partial derivatives may be cumbersome. In such cases we can resort to perturbation approach, i.e., by introducing small variations. Suppose the primary variables involved are x_1, x_2, ... x_n, we can perturb the variables by Δx_1, Δx_2, etc. Using finite difference method, for small-enough values of Δx, the partia derivatives can be approximated by

$$\frac{\partial R}{\partial x_i} = \frac{R\left(x_i + \Delta x_i\right) - R\left(x_i\right)}{\Delta x_i}$$

Suppose there are two measurements x_1 and x_2, and the final result is $Z = F(x_1, x_2)$ for some function of F. If x_1 is pertubed by δx_1, then Z will be perturbed by $\left(\frac{\partial F}{\partial x_1}\right)\partial x_1$. Similarly for perturbation of δx_2 in x_2, the effect of that in Z will be $\left(\frac{\partial F}{\partial x_2}\right)\partial x_2$. Combining these two as in the previous case,

$$\delta z = \left[\left(\frac{\partial F}{\partial x_1}\delta x_1\right)^2 + \left(\frac{\partial F}{\partial x_2}\delta x_2\right)^2 \right]^{\frac{1}{2}} \tag{1}$$

If $Z = x_1 + x_2$ then, $\left(\frac{\partial F}{\partial x_1}\right) = 1$ and $\left(\frac{\partial F}{\partial x_2}\right) = 1$. It can be noticed that this gives the same result as before. Similarly if $Z = x_1 - x_2$, then $\left(\frac{\partial F}{\partial x_1}\right) = 1$ and $\left(\frac{\partial F}{\partial x_2}\right) = -1$, which also gives te sae resut.

If $Z = (x_1)(x_2)$ then $\left(\frac{\partial F}{\partial x_1}\right) = x_2$ and $\left(\frac{\partial F}{\partial x_2}\right) = x_1$

$$\delta z = \sqrt{x_2^2 \left(\Delta x_1\right)^2 + x_1^2 \left(\Delta x_2\right)^2} \tag{2}$$

Thus the fractional error is the square root of the sum of squares of the fractional errors in its parts. For example, if $x_1 = 120 \pm 8$ and $x_2 = 10 \pm 3$ andapplying the above equation we get,

$$(120 \pm 8)(10 \pm 3) = 1200 \pm 1200\left[\left(\frac{8}{120}\right)^2 + \left(\frac{3}{10}\right)^2\right]^{\frac{1}{2}}$$

$$= 1200 \pm 368$$

11.5   STATISTICAL ANALYSIS OF ERRORS

11.5.1 Normality Assumption

Theoretically we should get the same value for a measurement when an experiment is repeated under the same conditions and environment but, in practice one can notice the variations in the measurements made. Suppose an experiment is repeated n times and we get he output variable of importance as $x_1, x_2, \ldots x_n$. All values are not equal. The mean value of the response or output variable for these n experiments will be $\sum x_i / n = \bar{x}$.

The difference between this mean value $\bar{x}$ and the measurement x_1, i.e., $\bar{x} - x_i$ is known as deviation or error. These errors are called random errors and are scattered around the mean value $\bar{x}$. By using the tools of statistical analysis we can estimate the uncertainty associated with the measurement.

The data collected from repeated experiments is known as sample data and the average of the sample ($\bar{x}$) is the best estimate of the true value of the individual measurements. If the number of repetitions (n) tends to infinity ($n \rightarrow \infty$), the average value represents the population mean. In many circumstances, it may not be possible for the experimenter to repeat experiments to collect as many data points as may be necessary to describe the underlying population. Authors like Holman, have suggested that at least 20 measurements are required in order to get reasonable estimates of standard deviation.

The uncertainty of a measured value (x_i) is an interval around the mean value ($\bar{x}$), such that any repetition of the measurement will produce a new result that lies within this interval. These uncertainties are always stated with probability and therefore, any measured value has the stated probability to lie within the confidence level.

There are several ways to indicate the error. The maximum error could be specified as $\Delta x_{max} = (x_{max} - x_{min})/2$ and no measurements should fall outside $\bar{x} \pm \Delta x_{max}$. The probable error Δx_{prob} specifies the range $\bar{x} \pm \Delta x_p$, which contains 50% of the measured values. Another indication of error is the average deviation which is the average of all deviations from the mean, i.e., $\Delta x_{av} = [\sum |x_i - \bar{x}|]/n$. Regarding measurement which follows Gaussian distribution, about 58% will lie within .

It is generally assumed that random errors are normally distributed. This is on account of the fact that many small errors contribute to the final errors and that each small error is of equal magnitude and equally likely to be positive or negative. Because of the law of large numbers and the fact that experiments are controlled, the assumption of normal distribution is valid for random errors. Therefore, a researcher can quote the error in terms of the standard deviation of the normal distribution fitted to the observed response data of the experiment. The probability of obtaining a value x in normal distribution is given by,

$$P(x) = \frac{1}{\sigma\sqrt{2\pi}} \exp\left[-\frac{1}{2}\left(\frac{x_i - \bar{x}}{\sigma}\right)^2\right]$$

where $\bar{x}$ is mean value and x_i is the ith observation. The maximum probability occurs at $x = \bar{x}$ and the value of probability is,

$$P(\bar{x}) = \frac{1}{\sigma\sqrt{2\pi}}$$

From the above equation, we can infer that smaller values of standard deviation produce maximum probability.

11.5.2 The Standard Error

It is a common practice to quote error in terms of standard deviation (σ) of the normal error distribution. The standard deviation is a measure of the spread of the distribution curve; the larger the value of σ, the flatter the curve and hence large expected error of all measurements as shown in Figure 11.1. It gives an idea of the reiability and precision of data collection.

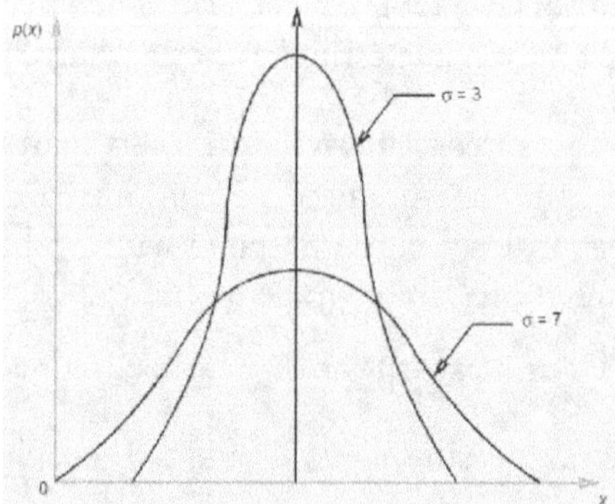

Figure 11.1   Error distribution

The standard deviation of any distribution is given by,

$$\sigma = \left[\frac{\sum \left(x_i - \bar{x} \right)^2}{n} \right]^{\frac{1}{2}}$$

The standard deviation is calculated based on n sample values and it may not be a true estimation of population standard deviation. Therefore for calculating the true standard deviation (σ_x) we use the relation,

$$\sigma_x = \left[\frac{\sum \left(x_i - \bar{x} \right)^2}{n-1} \right]^{\frac{1}{2}}$$

Here, we divide by $(n-1)$ and not by n to get the best estimate of σ. This is due to the fact that the value $(x_i - \bar{x})^2$ as calculated is always less than $(x_i - x_t)^2$ where x_t is the true value of the mean. In practice we do not know x_t and in probability theory it has been shown that this underestimate is corrected by using $(n - 1)$ instead of n. However, when comparisons are made with a known population or standard, it is proper to use n for the standard deviation, e.g., calibration of a pressure gauge against a known pressure.

11.5.3 Confidence Interval for Error

As the random error is normally distributed, it was stated that an analyst can fit the response data to a normal distribution for indicating errors in terms of standard deviation. In order to specify a probability for the credibility of the results, a confidence interval is fixed choosing an appropriate level of significance (α). The level of significance indicates the maximum probability of risk of rejecting the null hypothesis (H_0) when it is true. In other words, it indicates the probability with which a measurement on repetition of the experiment falls within the confidence interval. Some commonly usedconfidence level parameters are as follows:

Significance level (%)	Confidence level (%)	Confidence limits
1.0	99	$\pm 2.57\,\sigma$
5.0	95	$\pm 1.96\,\sigma$
10.0	90	$\pm 1.65\,\sigma$

For example, we expect that the mean value ($\bar{x}$) will lie within $\pm 2.5\sigma$ with less than 1 per cent error if the confidence level is 99 per cent. In this case the level of significance (α) is 1 per cent, which is (100 – 99 = 1).

If the number of times an experiment is repeated is less than 30, the normality assumption may not give satisfactory results. Therefore, in such cases the researcher can use the student's t-distribution. The confidence interval can be worked out as $\bar{x} \pm \delta$, where $\delta = t . \sigma/\sqrt{n}$. The value of t is obtained from the table of t-distribution knowing n and the level of significance ($\propto$). The example given below show how confidence interval is calculated using the t-distribution for an experiment.

Example An analyst made observations by repeating an experiment 10 times. If the values obtained are 1.20, 1.50, 1.68, 1.89, 0.95, 1.49, 1.58, 1.55, 0.50 and 1.09, indicate as to how the analyst will rport the error.

The ten observations (x_i's) are,

$$x_i\text{'s} = 1.20, 1.50, 1.68, 1.89, 0.95, 1.49, 1.58, 1.55, 0.50, 1.09$$

$$\sum_{i=1}^{10} x_i = 13.4 \qquad \therefore \bar{x} = \frac{13.4}{10} = 1.34$$

$$\sigma = \left[\frac{\sum \left(x_i - \bar{x} \right)^2}{n-1} \right]^{\frac{1}{2}} = \sqrt{\frac{1.497}{10.1}} = 0.17$$

The standard deviation is 0.17. As the sample size is less than 30 the analyst can assume, t-distribution with 9 degrees of freedom. The t-value for two-tailed test with 95 per cent confidence level and 9 degrees of freedom is, $t_{9,0.95} = 1.83$. Therefore the analyst can report that the cnfidence interval for the experiment result is,

$$\bar{x} \pm t_{9.95} \left(\sigma / \sqrt{n} \right)$$

$$= 1.34 \pm \frac{1.83(0.17)}{\sqrt{10}}$$

$$= 1.34 \pm 0.1$$

Any repetition of the experiment, the response measurement will fall within the interval of 1.24 to 1.44 for 95% of occasions.

11.5.4 Error Bars

The error bars are nothing but graphical representation of errors in the measurement made in an experiment. To statistically describe uncertainty or error in a measurement, analysts follow two approaches. One approach is using standard deviation (SD) of a single measurement often referred to as standard deviation and the other one is using standard deviation of the mean often called as "standard error" (SE). Since we represent the means in our graph, the standard error is the appropriate measurement to be used for the error bars. The standard error is calculated by dividing the standard deviation by the square root of the sample size n (ie., $s/\sqrt{n}$. From this, one can see that as the sample size n increases, the error tends to become smaller. Therefore our confidence in our mean value is increased when we make more measurements (i.e., sample size).

We can illustrate the use of error bars by means of an example.

An experimenter studied the performance of an I.C. engine by recording the power generated in kilowatts with change in the engine speed in RPM. Each trial was repeated fiv times and the results are tabulated as given below.

Trail	Engine speed in RPM			
	1200	1400	1800	2200
1	52	60	74	85
2	47	64	70	80
3	56	58	78	79
4	57	63	69	88
5	48	56	71	90
Σx_1	260	301	362	422
$\bar{x}$	52	60	72	84
s	4	3	3.3	4.3
SE	1.8	1.34	1.48	1.92

From the table of values shown above, we can specify that the power generated by the engine at 1800 RPM is 72 ± 1.48 kilowatts. This indicates that the mean power generated is 72 kW with a standard error (SE) of ± 1.48 kilowatts. The standard error (SE) can be superimposed on the graph. Speed in RPM vs power generated in kW are plotted as shown in the Figure 11.2.

The error bars shown in the above line graph represent a description of how confident we are that the mean indicates the true value of power generated at various speeds. In case we compare with the original scatter plot, we can notice that the original values of data are more widely scattered above and below the mean. If the error bars are wider then this gives less confidence about the particular value and if they are closely scattered then the width of error bars is less. Thus, the error bars give a general idea of how accurate a measurement is or conversely how far a reported value is from the true (error-free) value. Also the error bars can be used to show how good a set of data fits statistically with a given function. Many commercial software packages proide facilities for calculation and plot of error bars.

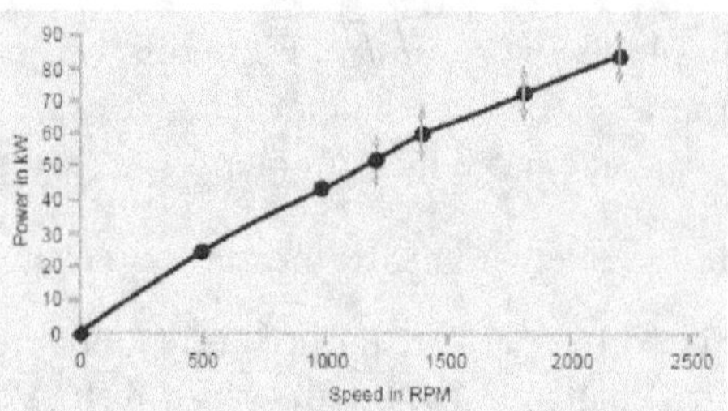

Figure 11.2   Representation of error bars

11.6   IDENTIFICATION OF LIMITATIONS

It is imperative on the part of researchers to spell out the limitations of the study they have carried out and also identify scope for further work while they do data analysis and interpretation. In particular the following aspects are worthy of consideration.

i.   The researcher must spell out the constraints of his/her experimental set-up, resulting in limitations ±1.48 kilowatts on the outcome of his/her study. The constraints may be due to resource availability or time frame. Therefore, the researcher should indicate as to how the outcome of the experiment can be improved if such constraints are addressed.

ii.   A model is developed for a specific purpose and the scope is mainly limited to the objective of the study. The researcher might have abstracted real-life situation for simplicity. Therefore, the researcher should explain as to how by improving the abstractions, the response of the model can be enhanced.

iii.   Based on error analysis, the researcher should identify the shortcomings in his/her study and how by increasing replications or otherwise the error could be reduced.

iv.   If there is a significant gap between the findings of the researcher and accepted theoretical outcome already available, the researcher should explain the inadequacies of his/her model or experiment as the case may be and how removal of such inadequacies will improve the result.

v.   Based on the outcome of the research work, if generalization could not be made, the limitations of the study, which are responsible for this must be highlighted.

vi.   Any relevant aspect or question the study has not addrssed, the reasons thereof should be explained properly.

12
Use of Optimization Techniques

12.1   CONCEPT OF OPTIMIZATION

Optimization is the act of obtaining the best result under given circumstances. An engineer has to make decisions at several stages—when designing, constructing and maintaining an engineering system. The ultimate goal of all such decisions is either to minimize the effort required or to maximize the desired benefit. Since, the effort required or the benefit desired in any practical situation can be expressed as a function of certain decision variables, optimization can be defined as the process of finding the conditions that give the maximum or minimum value of a function.

The optimum seeking methods are also known as mathematical programming techniques and are generally studied as a part of Operations Research (OR). Operations research is a branch of mathematics concerned with the application of scientific methods and techniques to the decision-making problems, with the objective of establishing the best or optimal solution(s). In the early days of development, the OR techniques were primarily used for resource optimization.

There are several ways the OR techniques can be classified. One such method of classifying OR techniques is i) mathematical programming techniques, ii) stochastic process techniques, and iii) statistical methods. Techniques like Linear programming, Non-linear programming, Dynamic programming, Network analysis, etc. form part of mathematical programming techniques. The stochastic process techniques include Markov process, queuing theory, simulation, reliability, etc. Techniques such as regression analysis, pattern recognition, design of experiments, factor analysis, etc. form part of statistical methods.

12.2   THE PROBLEM STATEMENT AND SOLUTION

12.2.1   Statement of an Optimization Problem

Constrained problem   An optimization problem can be stated as,

Find $X = \begin{bmatrix} x_1 \\ x_2 \\ x_3 \end{bmatrix}$ which minimize $f(x)$

Subject to the conditions or constraints,

$g_j(x) \le 0$, for $j = 1, 2 \dots m$

$h_j(x) = 0$, for $j = 1, 2 \dots p$

$x_i > 0$, for $i = 1, 2 \dots n$,

where,

X is a n dimensional vector called the "design vector" or "design factors",

$f(x)$ is the objective function,

$g_j(x)$ are inequality constraints and

$h_j(x)$ represent equality constraints.

It may be noted that the number of variables n and the number of constraints, m and/or p, need not be related in any way. The above problem is a constrained optimization problem. For the problem to be valid the functions $f(x)$, $g_j(x)$ and $h_j(x)$ must depend upon some or all of the input variables. The number of independent equality constraints must be less than or at the most equal to the number of input variables (i.e., $p < n$). If all the functions $f(x)$, $g_j(x)$ and $h_j(x)$ are linear in input variables, the problem can be called a linear programming problem. If any of these functions is non-linear, then the problem will be referred to as non-linear programming problem (NLP).

Unconstrained problem   In the case of unconstrained optimization problem, it will have only the objective function with no constraints. The problem can be stted as,

Find $X = \begin{bmatrix} x_1 \\ x_2 \\ \cdot \\ \cdot \\ \cdot \\ x_n \end{bmatrix}$ which minimize $f(x)$

where $f(x)$ is the objective function, X is an n dimensional vector and $x_i > 0$ for $i = 1, 2 \dots n$.

12.2.2 The Optimal Solution Point

In an optimization problem with only inequality constraints $g_j(x) < 0$; $j = 1, 2 \ldots m$, the set of values of x that satisfy the equation $g_j(x) = 0$ in the solution space is called a constraint surface. The constraint set for a problem is the collection of all feasible solutions. Mathematically S is a collection of points satisfying all the constraints,

$$S = \{\, x \mid g_j(x) = 0;\ j = 1, 2 \ldots m,\ h_j(x) = 0;\ j = 1, 2 \ldots p\,\}$$

Thus, S represents a set of feasible solutions and is called "feasible region". Feasible region shrinks when more constraints are added to the problem and the space expands when some constraints are removed. Likewise the locus of all points satisfying $f(x) = C$ (constant) forms a hyper-surface in the feasible space and for each value of C, there corresponds a different member of a family of surfaces. These surfaces are called "objective functionsurfaces".

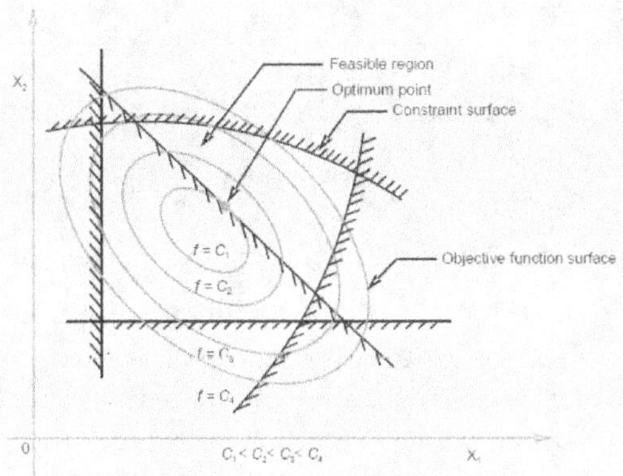

Figure 12.1   Search space

Once the objective function surfaces are drawn along with constraint surfaces, the optimum point can be determined without much difficulty as shown in the Figure 12.1 for two input variables case. However, once the number of variables exceeds two, the constraint and objective function surface become complex and it is very difficult for visualization. We may have to solve the problem purely as a mathematical problem.

12.3   OPTIMIZATION TECHNIQUES

12.3.1 Classification of Optimization Techniques

In order to transcribe a verbal statement of the given problem into a mathematical formulation, the following three steps are necessary.

- Identify and define input variables (factors).

- Specify the objective function and express it in terms of input variables.

- Identify constraints and develop expressions for each constraint in terms of input variables.

It must be recognized that the overall process of designing system in different fields of engineering or purpose is roughly the same. Perhaps, the statement of the problem can contain terminology specific to the domain under consideration. Once the problems of different fields are stated using standard notation, they all look alike.

The optimization techniques can generally be classified as indicated i Figure 12.2.

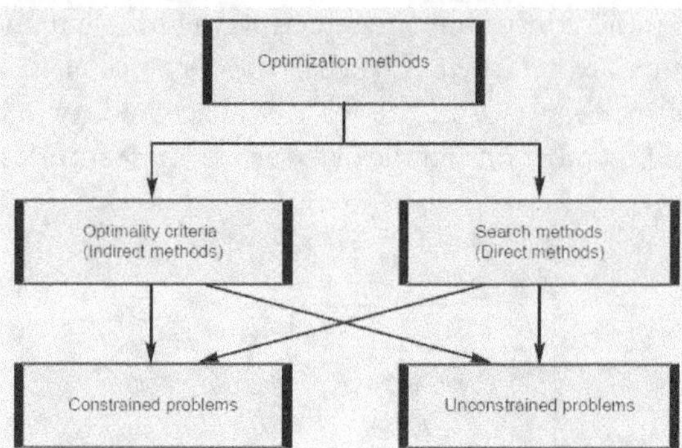

Figure 12.2 Classification of optimization techniques

Optimality criteria are the conditions that a function must satisfy at its minimum or maximum point. Techniques seeking solutions to optimality conditions are often called indirect methods. In these methods we assume that the functions are differentiable twice with respect to the input variable and the derivatives are continuous. We can use differential calculus to find the unconstrained maxima or minima of the given problem. These methods are often referred to as "classical methods".

The direct or search techniques are based on a different philosophy. Here, we start with an estimate of the optimum solution, which satisfies the optimality criteria, thereafter, we improve it iteratively searching the feasible space for optimum points. Once the difference between two successive improvements is very marginal or negligible, we terminate the iteration. Several search techniques like exhaustive search, interval halving, Fibonacci method and golden section methods are available for one-dimensional search, i.e., problem involving only one independent variable.

12.3.2 Scope of Optimization Technique

In many research projects, the researcher may be interested to find the value of levels of independent variables (factors) that yield the optimum (maximum or minimum) values of the response or dependent variable. If the dependent and independent variables are both quantitative and continuous, one can use Response Surface Methodology (RSM) adopting steepest ascent (or descent) method. Experimental studies provide the value of response variable only for the given set of input variables assigned by the researcher. From the response for the limited number of experimental results, the researcher decides which is the best among them for the assumed conditions. As it depends upon the selection of input variables, it is difficult to accept whether they really represent optimum level of input factors absolutely.

It may so happen in certain studies that the researcher may suspect multiple peaks in the response surface or there may be several saddle points. These may pose more difficulty in searching for a global optimum. The presence of multiple peaks in the response surface may be due to the occurrence of some fundamental changes in the response of the system under certain combination of input factors.

12.3.3 Traditional Methods

Many of the real-life problems involve multivariable and non-linear relationship. Several algorithms and solution techniques have been developed for the above problems. Some are general in nature and some solution methodologies are problem-specific. With the advent of computers many computer-based solutions for multivariable non-linear problems have been proposed. Of these, search techniques assume greater significance in computer-based solutions. Traditional search and optimization methods can be classified into two distinct groups—those requiring information about derivative values in determining search direction for optimization (indirect methods) and direct methods that rely solely on the evaluation of the objective function.

We first consider methods for unconstrained multivariable optimization. Newton's method is an indirect technique that employs a second-order approximation of the function. This method has very good convergence properties, but it can be an inefficient method because it requires the calculation of $n(n + 1)/2$ second-order derivatives, where n is the number of input variables. Therefore, methods that require the computation of only first derivatives and use information from the previous iterations to speed up convergence have been developed. The DPF (Davidon, Fletcher and Powell) method is one of the most powerful methods used. Updates and improvements on this method were provided by Broyden, Fletcher, Goldfarb and Shannon in what is called the BFGS method. It appears that this algorithm has better rating with practitioners.

One of the earliest direct search methods was the method of Hooke and Jeeves. Because it does not use derivatives, it is fast and is less sensitive to irregular or discontinuous functions. However, this method has problem in highly constrained situations where a search cannot progress further when the contour lines and the inequality constraint line have certain orientation. Other direct search techniques are the Simplex method (no relation to the technique in linear programming) and Powell's quadratic convergence method.

Optimization of non-linear problems with constraints is a more difficult area. There is a general approach to this class of problems and is called the Generalized Reduced Gradient method (GRG). The idea of GRG is to convert the constrained problem into an unconstrained one by direct substitution. However, with non-linear constraint equations, this is not feasible by direct substitution and the procedures of constrained variation and Lagrange multiplier must be used.

Another approach that can be adopted is successive linearization of the constraints and the objective function of a non-linear problem and solving them using the technique of linear programming. Sequential linear programming (SLP) is one such method. The recent and perhaps the best method of solving non-linear problem is sequential quadratic programming (SQP). Computer software solutions like GINO, NLP solver, and OPTISOLVE are available to carry out non-linear optimization on microcomputers.

Common difficulties with traditional methods   Because of the nonlinearities and complex interactions among variables that often exist in optimization problems, the search space may have multiple optimal solutions. In fact, many of them are local optimal solutions (sub-optimal) having inferior objective function values. While solving these problems using traditional methods, it may end up in local optimal solutions and no further movement is possible. Some of the common difficulties with most of the traditional direct and gradient-based techniques are the following.

- Choice of convergence to an optimal solution has greater dependence on initial solution.
- Most algorithms have the tendency to get stuck to a sub-optimal solution.
- An algorithm efficient in solving one optimization problem may not be efficient in solving a different problem, in other words it is context-dependent.
- Algorithms are not efficient in handling problems having discrete variables.
- Many algorithms cannot be efficiently used on parallel computing machine.
- As many algorithms are problem-specific, a designer has to be conversant with many optimization algorithms.

The above discussion suggests that traditional methods are not good candidates for an efficient search and optimization algorithm. Evolutionary optimization techniques like genetic algorithm, and particle swarm optimization (PSO) can be applied to a wide variety of problems.

12.3.4 Evolutionary Optimization Techniques

Evolutionary Algorithms (EA) are heuristics that use natural selection as their search procedure to solve problems. They have now gained immense popularity in real-world engineering search and optimization problems. To be economical and competitive, it is essential that we choose an optimal solution out of many feasible solutions satisfying the required performance criterion such as cost, performance, etc. One can notice that the last couple of decades witnessed constant development of new algorithms, theoretical achievements and novel applications.

Evolutionary algorithms (EA) are computerized search and optimization methods that work very similar to the principles of natural evolution. By using EA's intelligent search procedure we can find the best and the fittest design solutions, which may be difficult to find using other techniques. EAs are thus attractive and easy to use and they are likely to find the globally best solution, which is superior to any other solution. They are used in fields such as design of components, process design, pattern recognition, VLSI design, control systems, etc.

12.4 GENETIC ALGORITHM (GA)

12.4.1 Working Principle

In general many optimization problems in engineering are characterized by mixed continuous–discrete and continuous and non-convex search spaces. The conventional techniques are not only inefficient but also computationally difficult in such cases. Genetic algorithm is an evolutionary optimization technique, in which we mimic the evolutionary principles and chromosomal processing in natural genetics to derive optimal solution. It is a search and optimization technique developed by Professor John Holland of University of Michigan in 1965. Over the last one decade and more, the GAs have been successfully applied to a wide variety of problems, because of their simplicity, global perspective and inherent parallel processing.

Genetic algorithms are quite efficient in manoeuvring through large search spaces and identify optimal combination of solution, which is generally difficult to locate. In contrast to starting with the single-point solution of traditional optimization methods, in genetic algorithm we use a population of solution points (solution space). This means GA processes a number of solutions at the same time, i.e., in parallel. The solution space is converted into genetic space in GA by suitable coding of variables. In fact, the coding discretizes the search space even though the function may be continuous. While using genetic algorithm we have to consider the following aspects.

    i.   Definition of objective function

   ii.   Definition and implementation of genetic representation

  iii.   Definition and implementation of genetic operators

In GA we begin the search with a random set of solutions usually coded using proper structure. Every solution is assigned a fitness which is directly related to the objective function of the search and optimization problem under consideration. Thereafter, we modify the population of solutions to a new population by applying three operators similar to natural genetic operators, viz., reproduction, cross over, and mutation. We successively apply these three operators for each generation till a termination criterion is satisfied, thereby the optimal solution is achieved. A flowchart (Figure 12.3) is given to explain the working principle.

12.4.2 Solution Space

As far as general optimization methods are concerned, we move towards an optimal solution, which is the best among the available solutions. In GA we refer to the space containing all feasible solutions as "search space". Each solution can be marked by its value of the fitness of the problem. When we say that we look for a solution, it is implied that we are looking for an extrema (either maximum or minimum) in the search space. We have to recognize that the genetic algorithm is inspired by Darwin's theory of "survival of the fittest". The set of solutions represented by chromosomes are called populations. Solutions for one population are taken and used to form a new poplation (offspring) by selecting the fittest ones.

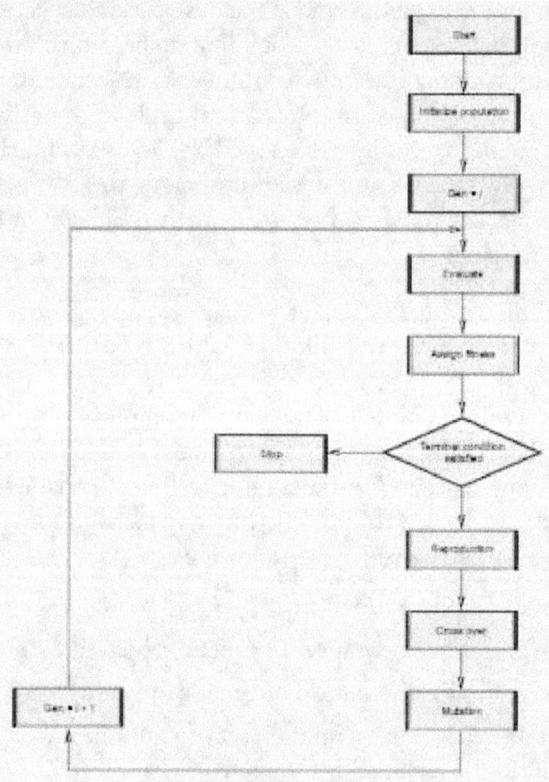

Figure 12.3 Flowchart of GA procedure

12.4.3 Objective Function

A genetic algorithm can be used to solve both unconstrained and constrained optimization problems. In the case of maximization prolem the objective function can be,

Maximize $f(x)$

$$x_i^{(L)} \leq x_i \leq x_i^{(U)} \text{ for } i = 1, 2 \dots n$$

Suppose we want to minimize $f(x)$ for $f(x) > 0$ then, we can write the objective function as

Minimize $\dfrac{1}{[1 + f(x)]}$

In case if $f(x) > 0$, instead of minimizing $f(x)$, we can maximize $\{-f(x)\}$. Thus, GA can handle both maximization and minimization problems.

12.4.4 Encoding of Solutions

In GA we start with random population of solutions and each solution has to be represented by a proper code. There are several ways of encoding, such as binary, octal, hexadecimal, permutation, etc. Binary encodings (i.e., bit strings) are the most common system of encoding due to various reasons. Most GA use fixed-length, fixed-order bit strings to encode candidate soutions such as x_1, x_2, x_3, x_4, etc. as given below:

$$\underbrace{11010}_{x_1}, \underbrace{1001001}_{x_2}, \underbrace{010}_{x_3}, \underbrace{0010}_{x_4}$$

Each binary-coded variable represents a chromosome and the length of the string is usually determined according to the desired accuracy of solution. By decoding we can get equivalent integer value for each string. For instance, binary code for the string 0101 is decoded as 5.

12.4.5 Fitness Function

The genetic algorithm mimics the theory of survival of the fittest and principle of nature in conducting the search process. Some of the genetic operators require the fitness function to be non-negative, while some others do not require this. Each string created in GA either in the initial population or in the subsequent generations is assigned a fitness value, which is related to the objective function value. For the maximization problem, we can use the fact that the string's fitness is equal to the string's objective function. For minimization problems the goal is to find a solution having the minimum value for the objective function and hence we can calculate the fitness function as the reciprocal of the objective function value. This ensures that the solution with smaller objective function value gets larger fitness.

The following are some useful transformations for fitness functions where $F(x)$ reresents fitness.

$F(x) = f(x)$ for maximization problems

$$F(x) = \frac{1}{f(x)} \text{ for minimization problems, if } f(x) \neq 0$$

$$F(x) = \frac{1}{[1 + f(x)]}; \text{ if } f(x) = 0$$

12.4.6 Genetic Operators

Reproduction   Reproduction (selection) is usually the first operator applied on a population. In reproduction we select good strings in a population and form a mating pool. The selected strings from the population form parents to cross over and produce offspring. According to the survival of the fittest theory, the best ones survive and create new offsprings. A number of reproduction operators exist in GA. The various methods used for selection include proportionate selection, Boltzmann selection, tournament selection, rank selection and steady-state selection. The essential idea in all the selection methods is that the above-average strings are picked up from the current population and duplicate of these are inserted in the mating pool.

A commonly used method is the proportionate selection where a string in the current population is selected with probability proportionl to the string's fitness. Thus the jth string is selected as $f_j \Big/ \sum_{j=1}^{n} f_j$. One way to achieve this proportionate selection is to use a roulette wheel with the circumference of the wheel marked for each string proportionate to the string's fitness. The roulette wheel is spun n times and a string is selected based on the pointer indication. The proportionate selection is also known as "Roulette Wheel selection". Instead of roulette wheel one can use random numbers also for the selection of strings.

Cross over   The cross over operator is applied next to the strings of the mating pool. A number of cross overs exist such as single-point cross over, multipoint cross over, etc. However, in almost all cross over operators, two strings are picked from the mating pool at random and some portion of the strings are exchanged between the strings as indicated below. It may be noted that in the example shown below the 3rd, 4th and 5th digit of oth the parents are interchanged thereby producing two children.

Parent 1	0 0	0 0 0	0 0 1 1 1 (child 1)
		$\Rightarrow$	
Parent 2	1 1	1 1 1	1 1 0 0 0 (child 2)

The main purpose of the cross over operators is to search the solution space so as to create a new population. The other aim is to perform the search in a way to preserve the information stored in the parent string maximally, because these parent strings are instances of good strings selected using the reproduction operator.

Mutation   Cross over operator is mainly responsible for the search aspect of genetic algorithm, but however the mutation operator is also used to supplement the purpose sparingly. Mutation involves flipping of a bit in the string with a small mutation probability, p_m. The mutation operator introduces new genetic structures in the population by randomly modifying the strings. The mutation operator changes a 1 into a 0 and vice versa of the chosen string, thus a new string is created. For example in the mutation operation changing a 0 0 0 0 0 into 0 0 0 1 0, the fourth gene has been changed by flipping its value, thereby creating a new solution. The need for mutation is to maintain diversity in the population. Simple GA uses population size of 30 to 200 with mutation rate from 0.001 to 0.5.

Genetic algorithm is an iterative optimization procedure. Instead of working with a single solution in each iteration, it works with a number of solutions simultaneously, (collectively known as populations) in each iteration. In every iteration (generation) a new set of strings (population) is produced using genetic operators, reproduction, cross over, and mutation and they are applied to the whole population. Further for every iteration the maximum or minimum value of the fitness function depending upon the nature of the problem is tracked.

In order to terminate the execution of a GA, we have to specify a criteria for stopping. We can terminate after a fixed number of generations or after a string with a high fitness value is located or all the strings in the population have attained a degree of homogeneity (i.e., a large number of bits have identical bits at most positions). Once we ensure convergence, the string values corresponding to the optimum fitness function value give the parameter of interest for the problem considered.

Many researchers are using genetic algorithms today as a means of optimizing complete problems (i.e., problems with many relative optima, but only one absolute minimum) with minimal effort. Application of pure numeric methods may not produce desired results. This is on account of search of a very small and limited space, which requires enormous effort. Further, there is no guarantee that the absolute optimum lies within the area scanned. In order to understand the concepts involved better, solving a simple problem using genetic algorithm is illustrated here.

Example   Find the optimal value (maximum) of the function, $f(x) = 26.12 + 0.714x - 0.119x^2 + 0.004x^3$ using genetic algorithm. Also check with the maximum value obtained using calculus approach.

The given function is, $f(x) = 0.004x^3 - 0.119x^2 + 0.714x + 26.12$. This function can be used as a fitness function for this problem since the problem is one of maximization.

Iteration I

i. Let us select the size of the initial population as 4 (i.e., 4 strings) with 5 digits in each string. The strings are obtained by selecting 20 single-digit random numbers consecutively and assigning binary digit 0 for the odd number and 1 for the ven number. The strings thus formed are given below with their encode.

String	Encode
1	10010
2	01001
3	10111
4	00010

ii. The values of the fitness function are calculated using he decoded value (x) and their percentage is given in the table below.

String	Encode	Decoded value (x)	Function value $f(x)$	Percentage (%)
1	10010	18	23.74	22.62
2	01001	9	25.85	24.63
3	10111	23	28.26	26.92
4	00010	2	27.11	25.83
Total			104.96	100.00

iii. Using the concept of survival of fittest we have to select the strings having highest function value or percentage for the second generation. For this we adopt proportionate selection method. We can use roulette wheel or random numbers to selec. The strings selected and the formation of new strings is as follows:

String No.	String	Selection order	New String
1	10010	–	00010
2	01001	(iv)	10111
3	10111	(ii) & (iii)	10111
4	00010	(i)	01001

iv. The cross over operation is made by using random numbers. In the instant case 1st pair is 1 and 3 and second pair is 2 and 4. The cross over site is the third digit. The cross over opeation and the new string formed for the next iteration is given below:

1st pair

$$
\begin{array}{ll}
000|10 \\
101|11
\end{array} \implies
\begin{array}{l}
00011 \\
10110
\end{array}
$$

2nd pair

$$
\begin{array}{ll}
101|11 \\
010|01
\end{array} \implies
\begin{array}{l}
10101 \\
01011
\end{array}
$$

Iteration II

The steps of iteration I are repeatd for the second iteration. The results are tabulated as given below:

String	Encode	Decoded value (x)	Function value $f(x)$	Percentage (%)
1	10011	3	27.30	26.08
2	10110	22	26.82	25.62
3	10101	21	25.67	24.52
4	01011	11	24.89	23.78
Total			104.68	100.00

After the second iteration (generation) the maximum value of the function is 27.30 and the corresponding string is 0 0 0 1 1 with a decoded value of 3. By the calculus approach to find the maximum value of the function, we take the first derivative and equate it to 0 and then solve the equation to get the value for x, on substituting it in the main function f(x), we get the optimum value as 27.33. The discrepancy between the two values is only 0.03.

The case example given below illustrates the robustness of genetic algorithm in the optimization of parameters involved in a gear train design for automobile application.

Case example 1

Gears are generally used in many machinery and in automobiles to meet their varied power transmission requirements. Design of gears is rather complicated as it involves several parameters to be taken into account. The instant case deals with the design of spur gear with the design objective of maximizing the power output from the gear train for a given input torque. Various aspects considered in the design include bending stress, contact stress, plastic deformation, water loads, static and dynamic loads and interference.

The design variables (factors) taken into account in this study are, transmission ratio, module, face width and the number of teeth on the pinion. As high as 13 constraints were used in the problem to take care of compressive stress, bending stress and other aspects. For the Genetic Algorithm in this study the parameters used are the following.

Number of variables	:	4
Chromosome length	:	24
Cross over probability	:	0.8
Mutation probability	:	0.01
Cross over site	:	2
Population size	:	50

By employing GA in solving the optimization problem, there has been an improvement of 15.9% power transmitted (torque) over the previous power transmission capability. Though the actual procedure has not been detailed here, the purpose of this case is to highlight how a researcher can use the technique of Genetic Algorithm in design situations where optimization is needed.

Case example 2

This example illustrates how the performance of Genetic algorithm can be improved and the computational time be reduced. New strategies are devised to improve the fitness of parent solutions. As a part of the research study a researcher proposed three improved Genetic Algorithms, viz., ordered replacement GA, improved tournament GA, and league GA. The researcher not only developed them but also tested them for their efficacy.

In the ordered replacement GA, the researcher improved the fitness of the parent solutions in the current generation before creating the next generation. The offspring solutions were created only from those solutions which have fitness value higher than the average fitness value of generation. By keeping the average fitness value of the generation as the cut-off value, solutions having lower fitness values have been replaced by offspring solutions having higher fitness value.

Regarding the improved tournament GA, the researcher adopted an approach of using two better solutions selected deterministically to proceed for further reproduction, instead of merely using two parent and offspring solution as normally done.

In the league GA, league groups were selected and each solution in the league group was permitted to perform genetic operation with all other solutions in that league group. The best solution was selected deterministically from two parent and their offspring solutions. From the best solutions produced. "league-winning size" number of solutions were selected for the next generation.

 The researcher implemented the improved GAs in binary and real code. In binary-coded GA, multipoint cross over and bit-by-bit mutation performed. In real-coded GA, simulated binary cross over and polynomial mutation were used. Constraints have been handled based on the method of distinguishing the feasible and infeasible solutions and by penalizing the infeasible solutions by an amount of their constraint violations.

All the improvements proposed in these three GAs were tested using test problems. The researcher compared the results produced by the improved GAs, employing the test problems with best known values available in the literature. It was found that the improvements shown justified the effectiveness of improvements proposed.

These improved GAs were used for the design optimization of final drive in automobile. The aspects considered include failure factors such as bending stress, contact stress, face contact ratio, overloading and road conditions. Objective function of the study is to minimize the volume of the final drive. The design factors considered are transverse outer module, face width, number of pinion teeth and number of gear teeth. Optimal values for design factors were established using the improved GAs which satisfy all constraints considered.

This example is mainly included to illustrate as how a researcher can improve an existing Genetic Algorithm. Being a developing area, the scope for further research in Genetic Algorithm appears to be good.

12.5 PARTICLE SWARM OPTIMIZATION (PSO)

12.5.1 Basic Concepts

Most of the population-based search techniques are motivated by evolution as seen in nature. Some well-known examples are i) Genetic algorithm, ii) Evolutionary programming, iii) Evolutionary strategies, iv) Genetic programming. Particle swarm optimization (PSO) is one of the optimization techniques and a kind of evolutionary computation technique. This method is extensively used in solving problems featuring non-linearity and non-differentiality, multi-optimization, etc. All these techniques work in the same way, that is, updating the population of individuals by applying some kind of operators according to the fitness information from the environment. This enables the individual of the population to move towards a better solution.

The technique of particle swarm optimization has been developed taking inspiration from biological systems. The basic PSO is developed from search on swarms such as fish schooling and bird flocking. The assumption is that every information is shared inside flocking.

Consider a scenario where a group of birds are randomly searching for food in an area. It is assumed that there is only one piece of food in the area being searched. All the birds do not know where the food is but they know how far the food is in each iteration. So, the best and effective strategy is to follow the bird which is nearest to the food. PSO learned from the scenario and used it to solve optimization problems.

Particle swarm optimization is a stochastic optimization technique developed by Dr. Eberhart and Dr. Kennedy in 1995, inspired by the social behaviour of bird flocking or fish schooling as stated in the previous para. There are two commonly used versions of PSO, viz., i) global version, and ii) local version. Global version is faster but might converge to local optimum for some problems. Local version is a bit slower but not easy to be trapped into local optimum. One can use global version to get quick result and use local version to refine the search.

12.5.2 Steps Involved

PSO search algorithm like GA is initialized with a population of random solutions, called "particles" (birds). Each particle in PSO flies through the search space with a velocity that is dynamically adjusted according to its own and its companion historical behaviours. Each particle keeps track of its coordinates in the problem space which are associated with the best solution (fitness) it has achieved so far. The fitness value is also stored and this value is called "p-best". Another "best" value that is tracked by the particle swarm optimizer is the best value, obtained so far by any particle in the neighbourhood of the particle which is called "l-best". When a particle takes all the population as its topological neighbours then the best value is a global best and is called "g-best".

The various steps involved in PSO are given below.

i. Each particle is considered as volume-less (a point) in the d dimensional search space. The ith particle is represented as $X_i = [X_{i1}, X_{i2}, X_{i3}, \dots X_{id}]$.

ii. The best previous (the position givin the best fitness value) of the ith particle is recorded and represented as $P_i = [P_{i1}, P_{i2}, P_{i3}, \dots P_{id}]$.

iii. The index of the best particle among all particles in the population is represented by the symbol $gbest_d$.

iv. The rate of position change (velocity) for particle i is represented by $V_i = [V_{i1}, V_{i2}, V_{i3}, \dots V_{id}]$.

v. The particles are manipulated according to the following equation

$$V_{i,m}^{(t+1)} = W.V_{i,m}^{(t)} + C_1 * rand(\ ) * \left(P_{i,m} - X_{i,m}^{(t+1)}\right) + C_2 * rand(\ ) * \left(gp_{i,m} - X_{i,m}^{(t)}\right) \quad (1)$$

$$X_{i,m}^{(t+1)} = X_{i,m}^{(t)} + V_{i,m}^{(t+1)} \quad (2)$$

$$i = 1, 2 \dots n; \ m = 1, 2 \dots d$$

where,

n is the number of particles in the group,

d is the dimension,

i is the pointer of iteration (generation),

$V_{i,m}^{(t+1)}$ is the velocity of particle i at an iteration t,

$V_d^{min} \leq V_{i,d}^{(t)} \leq V_d^{max}$,

$X_{i,d}^{(t)}$ is the current position of particle i at iteration p_i,

gp_i is the best previous position of the ith particle.

c_1 and c_2 are two acceleration constants known as learning factors, and rand( ) and rand( ) are two random functions in the range (0,1).

In effect the above equation updates the particle velocity and positions.

vi.　All particles in PSO are kept as members of the population through the course of the run (a run is defined as the total number of generations of the evolutionary algorithm prior to termination).

vii.　It is the velocity of the particle which is updated according to its own previous position and the previous best position of its companions. The particles fly with the updated velocities. In other words at each generation every particle in PSO can fly in a limited number of directions which are expected to be good to fly according to the group's experience. Particle velocities on each dimension are clamped to a maximum velocity V_{max}. If accelerations cause the velocity on dimension to excel V_{max}, it is limited to V_{max}. This parameter is set by the user.

viii.　In PSO a parameter called inertia weight (W) is used for balancing the global and local search and equation.

ix.　A large inertia weight facilitates a global search while a small inertia weight facilitates a local search. By linearly decreasing the inertial weight from a relatively large value to a small value through the course of the PSO run, the PSO tends to have more global search ability at the beginning of the run, while having more local search ability towards the end of the run.

Thus at each step, we change the velocity of each particle taking into account its "p-best" and "g-best" locations. Acceleration is weighted by a random term with separate random numbers being generated for acceleration towards "p-best" and "g-best" location. We repeat this procedure till the termination criteria are achieved thereby reaching the global optimum.

PSO is an extremely simple algorithm and it appears efficient for optimizing wide range of functions. As already stated it is highly dependent on stochastic process. It may be noted that the adjustment towards "p-best" and "g-best" by the particle swarm optimizer is analogous to the crossover operation in the genetic algorithm. However, the unique concept of PSO is the flying potential through hyperspace accelerating towards better solution. It appears that the success of particle swarm optimizer lies in the particle's tendency to hurtle past its target. In fact, the stochastic factor allows a thorough search of space between regions that have been found to be relatively good. The momentum effect caused by modifying the extent velocities rather than replacing them results in overshooting or exploration of unknown regions of the search space.

As already mentioned, we can use the PSO to solve many engineering problems as it is easier to use. The algorithm needs very few lines of computer coding and it requires only specification of the problem and a few parameters in order to solve it. The use of PSO is illustrated below with a case example wherein PSO was utilized to determine the optimal filter constant which in turn was employed to tune PID controller parameters.

Case Study

Control of liquid level plays a vital role in floatation plants especially in industries like chemical, power plants, water-treatment plants, etc. The level process is a non-linear one but it is a self-regulating process reaching steady state for different inputs. To effect control over the level process, several control techniques are in use. Among the various control techniques employed, Proportional Integral Derivative (PID) controllers are robust and are employed by many industries. Tuning the PID controller plays a major role in deciding its performance. In this study to design the parameters for PID controllers, an Internal Model Control (IMC) approach was conceptualized by approximating the transfer function using McClaurin's series. Here, the tuning parameter comprises finding the filter time constant () which plays a crucial role in deciding the performance of the output.

This study employs gain-scheduled PSO-based Internal Model Control (IMC) to determine the single parameter λ. An optimal value of the filter constant has been arrived at by using PSO technique. The block diagram for online tuning of λ using PSO for interna model control regarding the tank level control is shown in the following Figure.

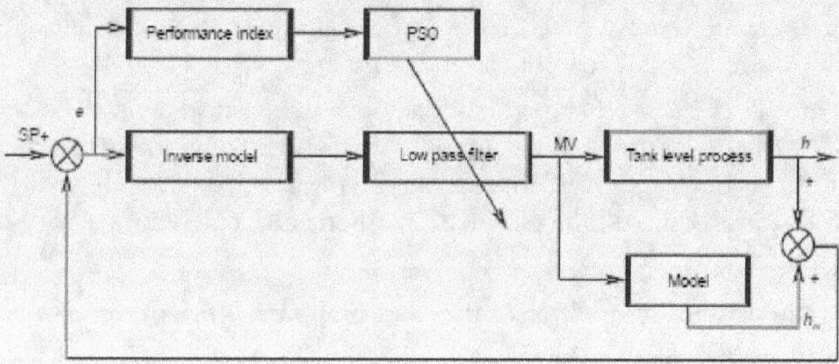

Block diagram for online tuning

For this study, the technique of PSO has been proposed as a method of tuning lambda (λ) which in turn generates the PID controller parameters. The technique uses PSO to tune the λ online for a minimum ISE for each region of five regions of the tank process separately The controller parameter is generated by minimizing the following error criteria.

$$ISE = \int_{o}^{T} \left[r(t) - y(t) \right]^{2} dt$$

where,

    $r(t)$ is the reference input

    $y(t)$ is the measured variable

At first, the position of particle, i.e., the value for λ is randomly initialized. The p_{best} and g_{best} are determined by the fitness function based on the inverse of the error criteria, ISE. The fitness function is defined as $1/(1 + ISE)$. Smaller the value of the fitness function, better is the performance of the system response with the specified PID parameters. Thus, the next movement for each particle can be computed by updating the velocity and position vector. After the required number of iterations, the optimal value for λ is arrived at. The number of iterations considered for the process simulation is 30 with an inertia of 1.0. Values for correction factors c_1 and c_2 adopted is 2.0 and 3.0 respectively and the swarm size fixed is 25. The initial value chosen for is 0.6268. A gain scheduler is used to control the operating conditions. The servo control of the tank process effected the set point changes at different sampling instants. The results indicated that the control system proposed is capable of tracking set points accurately and it is a solution for non-linear process. The novelty in this study is the optimal design of the IMC controller using PSO for global optimization of the parameter λ, which is an input to the process.

12.6 COMPARISON OF PSO WITH GA

There are many similarities between PSO and GA. A comparison of these techniques is given below.

- Both start with a group of randomly generated population.
- Both have fitness values to evaluate the population.
- Both update the population and search for the optimum with random techniques.
- Both systems do not guarantee success.
- PSO does not have genetic operators to update but, particles update themselves with the internal velocity.
- PSO also has memory which is important to the algorithm.
- Compared to GA, information sharing mechanism in PSO is significantly different. In GAs, chromosomes share information with each other. The whole population moves like a group towards optima area while in PSO, only "g-best" or "lbest" gives out information to others. It is a one-way information-sharing mechanism.

13
Writing of Papers and Synopsis

13.1 INTRODUCTION

Research scholars are expected to write and publish papers and articles for publication in reputed journals and magazines. Often they do this as a part of the requirement of the research programme, especially for doctoral programmes. In addition, the research study culminates in the preparation of the thesis based on the results of the study for evaluation by examiners. After evaluation, the research scholar has to defend his/her thesis in the viva-voce examination. All these require adequate knowledge and skill on the part of the research scholar in preparing technical report for each need. In fact even the brilliant findings of a research study are of little value unless they are effectively communicated to the concerned audience. This chapter deals with established practices, guidelines and rules which should be followed in the preparation of papers for journals and synopsis based on his/her research.

13.2 AUDIENCE ANALYSIS

13.2.1 The Basic Communication Model

Writing of a technical paper for publication in a journal or writing a research thesis for evaluation by examiners is an exercise in written communication. The basic communication model is given in Figure 13.1

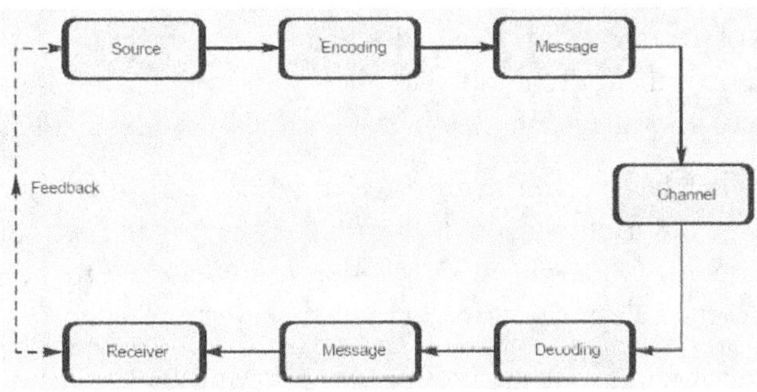

Figure 13.1   Basic communication model

The written communication can be in the form of papers, synopsis, brief memorandum reports, theses, technical proposals, etc. In terms of the communication model, the source is the mind of the researcher or writer. The process of encoding consists of translating the idea from the mind to words on paper. The channel is the pile of manuscript papers. Decoding the message depends upon the ability of the reader to understand the language and familiarity with the ideas presented in the report. Thus, the final receiver is the mind of the reader. Noise may be present in the form of poor writing mechanics, incomplete diagrams, incorrect references, etc. As, there is no direct feedback, the writer must anticipate the needs of the receiver and attempt to minimize the noise.

The prime principle of written communication is to understand the audience for whom the report is addressed so that the researcher can anticipate and fulfil their needs. The natural tendency for us is to write for ourselves rather than for someone else. We have to overcome this and therefore, we must actively force ourselves to analyse the audience and then tailor our presentation to their needs, not ours. This may not always be easy as the audience is diverse in nature or is not familiar to us. Nevertheless we must do the best we can in appraising their needs and capabilities.

13.2.2 Audience for Papers and Articles

The distinction between a paper and an article may not be obvious or universally defined, yet it is worth to find subtle differences between them. A technical paper is a document mainly prepared for submission to a technical society or organizers of seminars and conferences or to reputed publishers for publication or for presentation. Such papers are usually reviewed by a group of referees who recommend acceptance or rejection with or without modification. An article is usually considered to be at a lower technical level when compared with paper, that appears in a reputed journal or in a society's transactions and it is not subject to peer review.

The audience for a paper published in reputed journals comprises researchers, students, scientists, professionals engaged in academics and in industry, planners, etc. As already stated, the editorial board of these journals do arrange for peer review by experts in the respective fields before publication. Sometimes, it takes more than a year to get a paper published in a reputed journal. Moreover every journal has its own style in the matter of presentation, length and depth of treatment. It is imperative that the researcher must tune his/her paper to the requirements of the journal in which they want to publish. The researcher cannot assume too much of knowledge among the readers and make the text hard to understand.

As the researcher has worked in the area in which the paper is published and has thorough knowledge of various aspects dealt with, he/she may feel certain concepts are too elementary or axiomatic to be explained. One has to understand that even experts in the subject would welcome a clear exposition of ideas that are not well known in the discipline.

While calling for papers for publication, an Area Editor for the journal Operations Research relating to the area of manufacturing, service and supply chain management has stated as follows. "We seek well-written papers that span the full spectrum from theory to practice. Papers may be methodological contribution, contributed by developing and analysing novel models motivated by current industrial problems or phenomenon; or make an empirical contribution by conducting and analysing tests of hypotheses related to mathematical models, validating the behaviour of systems or algorithms, or reporting the results of major implementation of specific methodologies. In all cases, however, the contribution must clearly be significant, relevant, and conceptually sound and must meet the rigorous standards of the journal". This statement would enable the researcher to appreciate the requirements of a paper.

All these require proper understanding of the readership of the journal so that researchers can orient their papers in the proper direction.

13.2.3   Audience for Synopsis

Another important technical report that research scholars submit on fairly completing their research work is synopsis. This is a sort of summary report and is submitted to the doctoral or research advisory committee consisting of 3 or 4 experts in the field. Even though they have been periodically reviewing the work of the research scholar, synopsis is the document they go through to ensure that the work is completed and the researcher has made some original contribution for according approval to the scholar to write the thesis. Not only is a synopsis required for approval by the research advisory committee of the scholar, but also it is used by the university to get acceptance of the prospective examiner to evaluate the thesis of the scholar.

The researcher has to understand the requirements of the synopsis and what the audience looks for in that, so that he/she can meet the needs in the report. Synopsis is a brief summary of the research work carried out, indicating the problem taken up for study, methodology explained briefly, major findings and the researcher's original contribution. Many universities have prescribed guidelines for writing synopsis.

13.3   PREPARING PAPERS FOR JOURNALS

A researcher, who wants to write a paper for publication in a refereed journal, has to prepare first a list of journals that are most relevant to the area of focus of his/her research paper. From a list of international journals the researcher should shortlist two or three journals taking into account the reputation of the journal readership, periodicity of publication and the impact factor of the journal. Reference to recent issue of these journals will indicate the instructions and guidelines prescribed by the editors of these journals for preparation of articles. After considering all these aspects, the researcher should select the journal for sending his/her abstract of the paper to get their initial reaction or for sending the full paper as per the guidelines and instructions of the journal selected.

When researchers want to publish papers in a refereed journal based on their research work, they should not simply copy the relevant portion of their thesis and send it to the editor of the journal. They have to prune down by removing irrelevant portions and editing it so that it is in tune with the style and requirement of the journal. Most often the scholar will have to rewrite and remove the footnotes and other obstructions for easy reading. Even the prose style may have to be different than the one adopted for the thesis. A prudent approach would be to send an abstract to the editor first and get the confirmation of their interest in the topic and approval for the submission of the full paper.

On many occasions researchers publish papers in reputed journals based on their research work even before preparing their theses. Sometimes, this may be a requirement prior to submission of thesis. In that case they have to prepare the paper keeping in mind the need of the journal to which they are planning to send the paper. Many technical journals require authors to follow a style manual provided by the journal. These style manuals provide specifications about how papers for that journal are to be written.

An outline is very essential for preparing any paper for journal publication. A model outline is given below.

- Abstract
- Keywords
- Introduction
- Methodology
- Discussion of results
- Conclusions and findings
- Acknowledgement
- Tables
- Figures
- References

The Abstract section is the opening section of a paper which provides a brief summary of the paper. It is the most difficult part as it contains the essence of the entire work reported. As abstracts of journal papers are available through internet, it should provide enough motivation for the reader to read the main article after seeing the abstract. Briefly it describes the objective, the problem studied, and how the solution was arrived. The abstract should also highlight the major findings and its a contribution to literature. The abstract enables the busy reader to determine its worth. Ordinarily the length should not be more than 250 to 300 words.

Four or five words of specific relevance to the subject matter of the paper should be given as "Keywords" just below the abstract in a journal. The keywords selected should facilitate search for the article through indexing services.

The Introduction section gives a brief background to the problem and explains how it is important. The problem statement should not only be simple but also it should be interesting. One should not over-estimate the reader's familiarity with the topic at this stage. The introduction sets the stage in the same way as an abstract but in greater detail. It should cover aspects like justification for choosing the topic, specific methodology adopted, objectives and scope of the investigation. Any specific keywords or terms used, should be defined properly so that it makes it easy for the reader to understand. The scope briefly addresses what has been covered in the paper.

In the Methodology section, depending upon the nature of the study, the details should be covered. For example a study involving experiments includes details such as plan of the experiment, equipment used, metrics and/or techniques employed. The details provided should be adequate so that another researcher is able to repeat the experiment following the methodology stated. It is better for the researcher to follow the instructions given by the journal publisher as to the extent of details to be given. While the presentation of methods has to follow a logical sequence and description, routine techniques such as ANOVA may be mentioned without any description or reference.

The Results and Discussion part should cover how the results were deduced and inferences were made based on the analysis carried out. The data in the form of tables or figures may be placed at the end of the report and referred by proper numbering in the results section. If the tables are limited and there are only one or two figures, they can be included in the results and discussion section itself. Both the practices are prevailing. All arguments based on data are developed here. The discussion should include how the data addresses the problem taken up for research and its relation to other research and established scientific concepts. Results of the current research reported in the paper should be discussed by correlating with the findings of earlier research work. In case there are any limitations, not only should they be reported but also be justifiably defended.

The Conclusion section should cover in a concise form the major conclusions drawn from the study and interpretations made. In general, this section is the culmination of the research work and it is directly related to the objective of the research.

Sometimes small appendixes are used for mathematical developments; sample calculations, etc. that are not directly associated with the main body of the paper and if placed in the main body may impair the logical flow of thought.

Aspects regarding how to refer, etc., will be covered under the section, Writing of Thesis Report.

13.4 PREPARATION OF SYNOPSIS OF RESEARCH WORK

Synopses are required to be submitted three to four months prior to the submission of the doctoral thesis by the scholar by many universities. This is required to get the prior acceptance of the prospective examiner of the thesis by the university. Synopsis is a concise outline survey of the contents of the doctoral work. If a synopsis is written before a thesis is finalized it is quite likely that certain discrepancies bound to arise between the thesis and the synopsis. Because of this, many academicians and supervisors insist on preparing the draft thesis first and then write the synopsis for submission to the doctoral committee. This appears to be a justifiable view.

Many universities have prescribed detailed guidelines for the preparation of synopsis for the research scholars to follow. A typical framework is given below.

- Introduction
- Motivation
- Objective(s) and scope
- Description of the study
- Conclusions and contributions
- List of publications made
- References
- Proposed content of the thesis

It is better that the title of the synopsis be the same as that of the title of the thesis, otherwise confusion may arise later. One should devote attention to the selection of title as it reflects the major contribution of the research work. It should not be too long and it should be crisp and convey the meaning properly.

Under the Introduction section, the researcher briefly outlines the technological/engineering/scientific relevance or significance of the research work to be reported in the thesis. The introduction can cover at the most one page but preferably half a page. The scholar should be precise and include only the relevant background information. In addition, he can provide information on past works by citing appropriate references.

The research scholar should further develop on the background material provided in the introduction section under Motivation section. He should bring the state-of-the-art in the chosen area of research and must clearly indicate the existing drawbacks and why further research is required to eliminate the drawbacks. The importance of the research problem identified should be on the basis of current status. Further, the researcher should enumerate those technical challenges he/she has to address to solve the problems posed in the study. This will enhance the quality of the research work. The motivation section can be up to one page at the maximum or preferably half a page.

The next section to be covered is objective(s) and scope of the proposed research work. Here, the researcher should state clearly the questions for which answers are sought through the research work. The conceptual, analytical, experimental and/or methodological framework within which the exercise was carried out should be defined.

Again this section could be restricted to half page if possible, otherwise a maximum of one page used.

In the section on description of the research study, the researcher has to give brief but sufficient details regarding the research problem(s) and the solution methodologies adopted or developed by him/her. In case the study involves simulation and/or experiments, these should also be explained succinctly. This section should also briefly cover how the required data were collected, checked, analysed and the techniques used in the analysis. Rationale of assumptions and Validation of Results should also be explained properly but in brief. The research scholar should also explain as to how he/she interpreted the results. If necessary, the research scholar may introduce sub-sections appropriately to improve presentation. As this is the main body of the synopsis, a maximum of ten pages may be allotted for this.

Under the heading Conclusions and Contributions, the researcher must enumerate all major conclusions and findings. The consequential advantages of the work to be presented in the thesis should be clearly brought out, highlighting original contributions of the research work. In addition, the researcher should also give the limitations of his/her research work, especially when the results cannot be generalized. In case, there are no conclusions to be drawn then the contributions of his/he work should be listed. Perhaps, the title of the section can be changed to "Summary of the work".

It is imperative that the researcher should list the publications he/she made arising out of the research work in the next section captioned "list of publications made". Here, the complete details should be given using the standard format internationally adopted. Only the papers published or accepted for publication should be included. Even "papers under review" may be included but definitely not "papers under preparation". The listing sequence will be international journal publications first, followed by national journal articles, then International conference papers and lastly articles presented in national conferences. If the researcher has obtained any patents arising out of the research work, it can also be included in this section.

The next section is the Reference section. Here, the references are listed in the same order as they are referred to in the synopsis, ensuring that all references listed in this section are properly referred to in the text. It is better to restrict the number of references to ten or less. The order of reference can be i) books ii) patents iii) journal paper iv) papers presented and published in conference proceedings and v) web page. The specification for reference will be covered elaborately under thesis writing.

The last section of the Synopsis should cover the proposed contents of the thesis. In this section, the researcher has to provide the title of chapters and sections of the proposed thesis. A maximum of one or two pages may be allocated for this section.

Figures and tables must be embedded along with the synopsis. They should appear immediately after the first time the figure or table is cited in the text. Every figure and table should be numbered properly. The figure should be legible and enough information should be provided so that it is "self-contained". Legends are made legibly so that they correctly describe the axes of a graph. In the case of units, the research scholar should adopt the international system, SI units. A synopsis can have a maximum of 16 to 18 pages in A4 size paper.

A synopsis can be written using both present and past tense. When we express an established knowledge, it can be expressed in present tense. As far as the results of the work of previous researcher and experiment carried out by the research scholar are concerned, they should be in the past tense. The use of passive voice has been practised for a long period of time. However, some universities encourage the use of active voice in reporting, the reason being that the active voice is less verbose and unambiguous. As the synopsis is not a public document and not a part of the thesis, the instruction of the university/institutions has to be followed.

13.5 REFERENCE CITATION

13.5.1 Citation of references

It is customary in research papers and in synopses to cite the work of others published in journals or reported in books, monographs and other sources to support the researcher's point or to develop further on the previous findings. Therefore, it is essential to give credit to the works both published and unpublished which the researcher has used and quoted in his/her report. This is done by citing such works in the text and listing them under references at the end of the report.

References that are appropriate for citing in scientific or technical reporting may be grouped as primary literature and secondary literature. Research articles which report the results of original research published in peer-reviewed journals constitute primary literature. Theses, project reports, proceedings of conferences, textbooks, unpublished data, etc. constitute the secondary literature. Some peer-reviewed journals do not accept citation of secondary literature and other sources. Nevertheless, research reports published by government research agencies, technical books and edited books containing series of articles written by experts may be cited as reference in research theses.

13.5.2 Referencing Style

In citing references in the text of the paper or synopsis, there are many variations. Several journals and universities prescribe guidelines and standards for citing references in articles and in the synopsis. Though there are many referencing styles, most journals cite references in one of the three ways discussed below:

i. *Name and year system*   This system is popular one and used in many journal articles and synopsis writing. It is known as Harvard System of Referencing. Here the references are not numbered, instead they are referenced as "author and year" and included either inside a square bracket or parenthesis as shown in the following examples.

1. Yoshimura (1997) hypothesized that aluminium hydroxide ...

2. ... improving tracking resistance due to self cleaning action too (Yosimura et al., 1995 and Xidong et al., 1993) and ...

3. ... conductivity of composite material is reported by Meyer et al., (2003) and Gorur et al. (1998) ...

4. Dimensional analysis is a widely used technique to explain experimental data (Isaacson et al., 1975).

5. When there are two or more references by the same authors, the first one and second are indicated as follows ... Afridi et al. (2002a) ... Afridi et al. (2002b).

Certain journals do not encourage citing of secondary sources and they are not included in the references. Citations in the text for such sources should indicate the primary sources from which they were taken and the primary sources should be included under reference. For instance, ... Nerdfais (cited in Snurd, 1995) found in ...

The Snurd source should be cited in the references.

The advantage of the name and year referencing system is that references can be easily added or deleted.

ii. *Alphabet-Number system*   In this system, a particular material included in the list of references is cited in the text of the report by indicating the author and serial number of the reference. It is a modification of the name and year system. The following examples indicate the usage of this system.

1. Model building as an integral part of scientific enquiry has been stated succinctly by Rosenblueth and Wiener [25]...

2. Robert Shennon [23] raises same doubt about the reasonableness of usage of ...

3. Construction of obstruction meters has been standardized by American Society of Mechanical Engineers [1 and 2].

4. According to Naylor et al. [20], the problem with stochastic convergence is ...

Here, all the references cited in the text are listed under references at the end of the paper or synopsis in the alphabetical order of authors and they are serially numbered. The number referred to in the text and the serial number included under references should match.

iii. *Citation order system*   The citation order system contemplates simply citing the references by the number in the same order in which they appear in the paper or synopsis.

For example a reference first cited in the text will be given serial number 1 and the second cited reference 2, etc. Many readers may like this system because they can quickly refer to the References, should they so desire. May be this system is better for short papers but may not be a good system for long articles. Here, inclusion and deletion of references pose considerable difficulties to the researcher. Even if he/she deletes one reference or adds one more, the whole numbering order has to be changed. Some authorities do specify hierarchy of references such as books, monograph, journal articles, etc. in the order.

13.5.3 Using Quotations

In research reports, sometimes it may be necessary to quote certain extracts from literature verbatim to stress the researcher's viewpoint. The researcher would have collected several extracts while doing literature survey and review, but should choose the most appropriate ones to incorporate in the report depending on the context. He/she has to consider and decide when to quote, what to quote and how to quote in the report. While the use of quotes cannot be totally avoided, they should be used judiciously and sparingly.

The research scholar can think of using the quotations, only when he/she is of the view that the original words of the author are more appropriate and that he/she can neither improve it nor modify it. Moreover, this can be pursued only when the researcher feels that such a suitable quotation will supplement his/her point or wants to repute or analyse ideas expressed by the original author. However, when to quote and what to quote depends upon nature of the problem investigated and the judgement of the researcher.

There are certain established conventions to use quotations. The quotations have to be in the exact words of the author without interfering with punctuation, spelling, capitalization, etc. In fact, the accuracy, the verbatim is more important. If a quotation is long, the researcher can use the relevant portion of the quote, omitting certain portions.

For instance, if the researcher introduces a quotation at the beginning of a chapter, the format could be,

"Reality is neither in the world nor in our models, but in the process of working back and forth between world and model".

–R.A. Singer

If a quotation is introduced within a paragraph of a text, the format is,

Albert puts it this way: "The fear of wrong decision is frequently more compelling than the possible adverse consequences of no decision".

As a general guidance, if the length of a quotation is four or less lines then the quote can be embedded in the text line itself enclosed within quotation marks. In case the quote is five lines or more in length, it should be indented, blocked and single-spaced with the quotation marks removed.

13.5.4 Listing of References

A researcher must include in the list of references only those items which he/she has cited in the text of the paper or synopsis. The purpose of the list of references is to point out the specific location of an idea or piece of information in the original source. The format of the reference section considerably varies among journals and universities according to the style of citing references in the text. Therefore, a researcher must be sure of the format that is appropriate to adopt, taking into account the journal prescription and the university guidelines.

A frequently used format by many journals and universities is given below:

1. *For journal articles* Last name and initial(s) of author, [followed by last name and initial(s) of co-authors, if any]. Article title. Name of the journal. Year [Month and Day, if necessary and available] of publication; Volume number (Issue number); Inclusive page number. An example is given below.

> Ullman, D., Wood, S., and Craig, D. "The Importance of Drawing in The Mechanical Design Process", Computers and Graphics, Vol.14, No.2, 1990, pp. 263–274.

2. *For books* Last name and initial(s), [followed by last name and initial(s) of co-authors if any]. Title of the book. Place of publication: Publisher; Year of publication. Number of pages. An example is as follows.

> Dym, C.L., and Little, L., Engineering Design: A project based introduction, 2nd edn., New York, John Wiley, 2003.

As the size of the article will be less than 10 pages and the synopsis size about 18 to 20 pages, only relevant and most important references have to be cited in the text and listed under reference at the end of the paper or synopsis. In fact some universities even prescribe limit for number of references to be included. Therefore, it is imperative on the part of the researcher to be selective to refer only to those aspects which are needed to emphasize his/her point in those documents and list them under reference. There are other aspects such as footnotes, quotes, etc., which will be covered in the next chapter on thesis writing.

3. *IEEE format for referencing* The Institute of Electrical and Electronics Engineering (IEEE) is one of the reputed and the largest professional association, which dedicates itself for the advancement of technological innovations and excellence for the society's benefit. They publish large number of refereed journals, organize international conferences, develop technology standards for various products and processes and also involve in educational activities. The IEEE format of referencing and bibliography is one of the well-accepted standards by the scientific and technological community and followed by many publishers and authors all over the world.

The following examples indicate how books, journal papers, etc. can be listed in the reference section of a paper or a book.

a. *Book referred*

    i. C.L. Dym, Engineering Design: A synthesis of views. New York: Cambridge University Press, 1994.

    ii. S. Rajasekaran and G.A. Vijayalakshmi Pai, Neural Network, Fuzzy Logic, Genetic Algorithms. New Delhi: Prentice Hall of India Ltd., 2003.

    iii. J.E. Jenson, et al., Soil Science. San Francisco: W.H. Freeman, 1978.

b. *Journal articles*

    i. C.Clark, "The Utility of Statistics of Random Numbers," Journal of Operations Research., Vol.8, No.2, 20–25, Mar–Apr.1962.

    ii. J.T. Luxhol and P.H.K. Hansen, "Engineering Curriculum Reform at Alborg", Journal of Engineering Education., Vol.85, No.3, 83–88, 1996.

c. *Paper in Conference proceedings*

    P.H. Diananda, "Some Probability Limit Theorems with Statistical Applications," Proc. Cambridge Philosophical Society, Vol.49, 1953, pp. 239–46.

d. *Paper presented in a Conference/symposium*

    H.T. Mann, Herbert T., "Metabolism of 5-HT Inhibitor", presented at 2nd Optical Fiber Sensors Conference, Boston, MA, 1991.

e. Personal interview

    Personal interview with Troy Kimnet, Chief Metrologist, KVUE Channel 24, Austin, Tx, April 18, 1993.

f. *Handbook/databook*

    User's Guide: Microsoft Word.vers.5.0, Microsoft, 1991.

g. *Patents*

    J.K. Gustafsson, U.S. Patent 3960010, 1976, Analog–digital convertor for a resistance bridge.

h. *Dissertation or thesis*

    R. Rajaprabu, "Insulation Characteristics of Silicone and EPDM Polymeric Blends for High Voltage Applications,"Ph.D. dissertation, Anna University.

14

Thesis Writing: Mechanics

14.1 INTRODUCTION

A research study culminates in the preparation of a document, which is a formal communication report, written by the research scholar to a specific audience. This report is mainly for the purpose of evaluation by examiners duly nominated by the University in case the report is for the award of degrees like M.Phil., M.S. or Ph.D. Universities prescribe guidelines for preparation and submission of dissertation reports. Regardless of specific guidelines a thesis report generally includes motivation for study, methods adopted, data collection, analysis, interpretation and conclusions drawn highlighting the original contribution of the researcher in the study.

Apart from being an instrument of recording significant information and decision-making, a thesis report has an intrinsic value. It develops the power of discrimination and judgement besides inculcating the skills of logical thinking and systematic presentation. Further, a researcher's thesis is an interpretative and conclusion-oriented type of report based on his/her in-depth study. All these aspects are covered in detail in this chapter keeping in mind the needs of students pursuing M.Phil., M.S. and Ph.D. degree programmes.

14.2 AUDIENCE FOR THESIS REPORT

The thesis or dissertation is mainly prepared for review by the supervisor and the committee of experts supervising the work of the researcher and finally by the examiners. Once the thesis is approved and becomes a contribution to the literature in the concerned field, it will be referred by the scientific and technical community working in the same or related field and even by the general public. Therefore, a researcher has to keep in mind the expectation of the audience while preparing the report.

14.3 STEPS IN WRITING THE REPORT

As already stated writing a research report is a major component of the research project. The research task remains incomplete till the report is written and presented. There are five operations involved in writing a high-quality report, which can be easily remembered with the acronym POWER, where

P	Plan the writing
O	Outline the report
W	Writing the report
E	Editing the report
R	Rewriting the report

14.3.1   Plan of Writing

In the planning stage of the report we are concerned with assembling the data, analysing the data, drawing conclusions from the data analysis and organizing the report into various logical sections. During this phase we consider the various facets of the work and provide a logical blend of the material. The initial planning of the report should begin before the work is carried out. In that way the planning of the work and planning of the report are woven together, which facilitates actual writing operation.

The choice of a suitable title for the thesis should not be treated lightly. The need for a proper title was also stressed in the synopsis section. The researchers and other professionals mostly search for information based on the title and decide whether to pursue a particular reference. Therefore, we should try to come up with the most descriptive and concise title that we can. If possible the title should clarify whether the research work is theoretical, experimental or a computer simulation and whether it is original or a review of previous work. The title should aid the literature searcher easily.

The researcher should also decide about the proper choice of a word processor he/she is going to use. The word processor software is a very useful tool due to its ability to move sections of the report by "cut and paste", to insert or change words, to check spelling and grammar, and to increase the readability by changing the format or size of type or font, etc. Commercial word processors have gradually become bigger, slower, less reliable and more unwieldy to use as they acquire more features. A researcher needs a word processor which is easy to use and also allows typing of mathematical symbols and equations conveniently.

A high-quality software for the production of technical and scientific documentation is LaTeX. It is a powerful, elegant, reliable and quality typesetting system. It is not a word processor, instead it encourages authors not to worry too much about the appearance of their document but to concentrate on getting the right content. Many scientists, engineers, economists and others use this software extensively for documentation of their articles, theses and other reports. Some universities have even prescribed this as a standard software for preparing the thesis for the Ph.D. programme. This software is freely downloadable from the website http://wwwlatexproject.org/

As a part of planning, the researcher should also decide as to how he/she is going to organize the files. One word processor file can be opened for each chapter and one for references. Notes can be kept in these files, as well as text. This gives lot of flexibility to swap between chapters if necessary. Apart from digital files, the researcher should have a physical filing system: a collection of folders with chapter numbers on them. All relevant materials collected from different sources can be put in the respective folders for easy reference as and when required, while writing the thesis.

It is advisable that a researcher should sit down with his/her supervisor and prepare a time-table for writing the thesis. The time-table should contain list of dates for submission of chapters to the supervisor for correction and comments. This will structure the time of the researcher and decide targets. If one aims to finish the whole thing merely by some distant date, they may prolong it more easily. If needed, the time-table can be prepared in the form of a bar chart.

14.3.2   Outline of the Report

An outline of the report essentially consists of a series of headings, sub-headings, sub-sub-headings, etc., which encompass the various sections of the report. Figures, titles (to indicate which results go where) and some other notes and comments may also be included. Once researchers prepare a list of chapters and under each chapter heading a reasonably complete list of things to be reported, it indicates that they have overcome the writer's mental block. When they sit down to type or write, their aim is no longer a thesis—a fearsome goal—but something simpler. The aim of the researcher now is to write a paragraph or section about one of the sub-headings. Perhaps, one can start to write with an easy one which gives self-confidence.

As a part of the preparation of an outline, a table of contents, which may be tentative, should be prepared. This is a framework like a skeleton of a human body to which the material is stuffed. The content should enumerate all chapters, main heading, sub-headings, etc. in a logical order. If main headings and sub-headings are numbered according to some scheme, it is better to keep such numbering system simple. Many universities favours decimal numbering system. For chapter number, Arabic numeral such as 1,2 are used with main heading numbered as 1.1, 1.2, etc. and sub-headings as 1.1.1, 1.1.2, etc. However, one should not go to the extent of numbering like 1.2.3.3.8, which is clumsy and confusing.

While formulating the chapter main headings, sub-headings, etc., the researcher should keep in mind the chapter outline that was formulated. This leads to a question as to how one can make an outline of a chapter and it requires a lot of cogent thinking and vision on the part of the researcher. A suggested approach is for the researcher to assemble all the figures prepared for the thesis and putting them in the order that the researcher would use it if he/she were to explain to someone what they all meant. Perhaps, the researcher can rehearse it if necessary. Once he/she finds a logical order, the keywords of the explanation may be noted down. These keywords will provide a skeleton for much of the chapter outline.

Once an outline is prepared, the researcher should better discuss with the supervisor. This step is an important one as it provides a chance to get useful suggestions and comments from the supervisor. Based on this, the researcher can refine the outline. A copy of the agreed outline should also be given to the supervisor for reference when reading chapters which may probably present out of order. If the researcher has a co-supervisor, the outline should be discussed also with the co-supervisor before finalizing it.

14.3.3 Writing the Report

This section forms the crux in writing a thesis. The various issues concerned here are,

- The extent of detail to be provided
- The method of presentation of the contribution of the researcher
- The issues involved in the mechanics of writing
- Presentation of graphs, charts, etc.
- Where to stop

Considering the issue of the extent of detail to be included, it should be more elaborate than for a scientific paper. The thesis is mainly read by thesis evaluators, it should be self-explanatory without any ambiguity. Also the thesis will be read by fellow researchers who are seriously doing research in the same area. They may be interested to find out exactly what the researcher of the thesis did. They may look for details of experimental arrangement, circuit diagram or a sub-routine of a computer program, or the mode of application of a particular technique and the context. Therefore the thesis should include workshop drawings, circuit diagrams and computer programs as appendices. The researcher who had written the programme knows each line of the code, but for somebody else to understand it properly, it should be extensively annotated. The researcher might have read other theses and so know the advantages of a clearly explained, explicit thesis and/ or the disadvantages of a vague one. Personal experience is the best judge in this.

It is incumbent on the part of the researcher to make a clear distinction between his/her views and that of others. If a researcher uses a result or observation from another source, the scientific literature wherein those results were reported must be indicated. The only exceptions are cases where every researcher in the field already knows it: dynamics equations need not be followed by a citation of Newton, circuit analysis does not need a reference to Kirchoff, and so on. The importance of this practice in science is that it enables the reader to verify the researcher's starting position.

In some vertical sciences like physics, results are built upon results like a building block. Therefore proper referencing allows the reader to check the foundations of the researcher's additions to the structure of knowledge in the discipline. It helps the reader to trace back to a level so as to judge its reliability. Good referencing also tells the reader, which parts are additions by the researcher to that knowledge. In a thesis written for the general reader who has little familiarity with the literature in the field, this should be especially clear. The researcher should avoid the practice of leaving out a reference in the hope that the reader will think that a nice idea or a nice bit of analysis, is that of the researcher. In today's digital world one can easily find out the truth.

14.3.4   When to End Writing the Thesis

The writing of a thesis should be carried out in the form of a rough draft using the maximum technical and computational skill at the command of the researcher. It should be in conformity with the table of contents prepared and the general outline adopted. However, the researcher should not be worried about perfection at this stage especially finer details such as punctuation, sentence structure, etc.

Fixing a deadline for completing the rough draft is very useful. The researcher should submit the thesis even if it seems that one more draft of a chapter, someone's comment on a particular section or some other refinement may be needed. One should not forget that a thesis is a very large work and that it cannot be made perfect in finite time. Inevitably there may be things that the writer could have done better. So, the researcher has to set a deadline, while preparing the time-table and has to strictly adhere to it.

14.3.5   Editing and Rewriting the Thesis

Editing is the process of reading the rough draft and employing self-criticism. It consists of strengthening the rough draft by analysing paragraph and sentence structure, economizing on words, checking spelling in general, asking oneself the question "why"? Editing can be the secret of good writing. It is better for the researchers to ask themselves embarrassing questions than to hear them from their technical readers, or supervisor.

As already mentioned elsewhere in this chapter, it is better for the researchers to give their rough draft to their closest colleagues who could take the trouble of going through the draft and offer comments, which can be considered for making amendments. The supervisor is bound to go through each and every chapter of the thesis, suggest corrections, additions and deletions. Sometimes the supervisor may even suggest rewriting certain portions of the thesis, or adding some graphs and tables.

The researcher might have published papers based on his/her work even before writing the thesis. The editorial board of the journal to which the paper is submitted normally conducts a peer review of the paper by experts. These experts offer critical comments, suggestions, etc., for improving the content, which the researcher carries out and sends the paper again for publication. When writing the thesis, all such comments received should be given full consideration. If by chance any omission took place, the researcher should carry out this while editing the thesis.

It is quite likely that the research advisory committee while reviewing the synopsis makes certain observations or comments. This not only calls for editing the synopsis but also for incorporating the suggestions made later in the thesis. With regard to editing, it has been said that the best writer makes good use of both the ends of the pencil.

After writing the rough draft, and before editing, it is generally good practice to take a small break, at least a day or two. This allows the researcher to forget the logical pattern used in writing the thesis and take the role of an unbiased reader while editing. Many mistakes or weak lines of thought that would normally escape unnoticed are thereby spotted. The rewriting operation consists of retyping or rewriting the edited rough draft to put it in a form suitable for the evaluators.

The researcher must be aware that how despite his/her reflection and editing many times, there will be something that could be improved upon. There is no point hoping that the examiners will not notice. Experience reveals that examiners do suggest some correction or addition of points since they meticulously go through the thesis taking time. Similarly during defence of the thesis in a public viva-voce, suggestions and comments are bound to arise, which the researcher has to incorporate in his/her thesis. Therefore editing and rewriting takes place at several stages of the research work.

14.4   MECHANICS OF WRITING

14.4.1   Style and Tense

The text of the thesis must be clear. Good grammar and thoughtful writing will make the thesis easier to read. Scientific writing has to be formal. Since scientific English is an international language, slang and informal writing will be difficult for non-native English speakers to understand. Therefore the researcher should use short, simple phrases and words rather than long ones. However, a researcher can use a complicated sentence to convey a complicated idea. The need is clarity in written communication.

Sometimes it is easier to present information and arguments as a series of numbered or bulleted points, rather than as one or more paragraphs. A list of points is usually easier to write. The researcher should be careful not to use this presentation too much. The thesis must be connected, and should contain convincing argument, not just a list of facts and observations.

An important stylistic choice is between the active and passive voice. The active voice ("He measured the temperature...") is simpler, and it makes clear what the researcher did and what was done by others. The passive voice ("the temperature was measured by him ...") makes it easier to write ungrammatical or awkward sentences. The researcher should be wary of dangling participles while using the passive voice. The choice whether to use active voice or passive voice is a question of taste. In scientific writing it used to be required that the author use the third person and passive voice. The idea was to make the writing sound scientific, i.e., objective and detached from any personal influences. Today many style manuals for high-level technical journals encourage a less formal and more conversational tone by using active voice.

Authors like Ernest O. Doebelin recommend the use of active voice and first person as the usual form unless there are specific requirements to the contrary. Some of the Australian universities also prefer active voice because it is clearer, more logical and makes attribution simple. The arguments against the use of active voice are that many theses are written in the passive voice and some very polite people find the use of "I" immodest. Considering all these it appears that the use of active or passive voice not only depends upon the choice of the researcher but also that of the university.

The choice of tense of verbs is often confusing to student writers. The following simple rules are usually employed by experienced writers.

i.   Past tense is used to describe the work done in a laboratory or in general to past events.

        Example   **Hardness readings were taken on all specimens.**

ii.   Present tense is used in preference to items and ideas in the report itself.

        Example   **It is clear from the Figure 4 that strain energy is the driving force for recovery).**

iii.   Future tense is used in making predication from the data that will be applicable in the future.

        Example   **The result given in Table 2 indicate that the environmental pollution will continue to increase in the next ten years).**

14.4.2   Organization of Sentences

An experienced reader expects an entire report or thesis to have a structure or organization. A conscious effort has to be made in the formation of proper structure of individual sentences. Information is interpreted more easily only when it appears in the form that most readers expect it to occur. This fact can be used to improve the sentences and paragraph structure.

A reader will generally expect the subject of a sentence to be followed fairly closely by the verb. When this does not happen, clarity suffers because, without verb, we don't know what the subject is doing and thus what the sentence is all about. Therefore, we should avoid separating the subject and verb too much and again this has to be checked when we proof-read the rough draft of the thesis.

Another aspect deals with the materials that we place in the topic position (beginning of the sentence) and the stress position (end of the sentence). The reader will naturally expect that material at the beginning of sentence will provide linkage (looking backward) to the previous sentence and context (looking forward) to the new sentence. Thus the topic position should be used for such transitional material which provides a smooth flow between sentences. Readers also expect a sentence to build emphasis at the end. Therefore, the researcher should strive to place the next important material in that location, the stress position.

The idea of the previous paragraph can be explained by means of an example to improve the understanding. Let the last two sentences of a paragraph be ... "A team member should spend many hours on some sub-set of the problem to develop ideas. Thereafter the team can discuss and consolidate the ideas developed by individual members". The words at the beginning of the last sentence, "team can discuss" are in the topic position while the stress position is "to consolidate the ideas". Suppose the opening sentence of the next paragraph is, "As a result of the consolidation of ideas, the team can evolve an integrated concept". Here, the phrase, "As a result of consolidation of ideas"... provides the linkage to the last sentence of the previous paragraph.

By suitable punctuation we can create stress positions, allowing several pieces of stress-worthy information to be clearly presented in a single sentence. This is most commonly made with a semicolon which creates essentially two sentences each with its own stress position. Such constructions lead to long sentences, which nevertheless are clear and easy to read. Some authors say that a sentence should not exceed 35 words. While lengthy sentences should not be the norm, length alone is not the only consideration. Even a 15-word sentence can be quite unintelligible while a properly designed 40-word sentence can be perfectly clear.

14.4.3   Paragraph Structure

The researcher should structure the paragraphs properly. Most writers prefer to begin a paragraph with a topic sentence. This sentence presents the idea to be developed in the rest of the paragraph. It is always better to have only one main idea in each paragraph; however simple ideas may require one or more sentences leading to evenly short paragraphs. The researcher should then try to treat several ideas in one paragraph by showing some relation among them.

Complicated ideas can lead to exceedingly long paragraphs. Long, unbroken sections of the text repel the reader's eye and overload the memory, so the researcher needs to force paragraphs in such cases, even while still developing the main idea. The researcher can usually achieve this by using suitable transitional words at the paragraph break.

It is better to have one or more paragraph breaks per page. The reader may recall the example quoted in the previous section as to how linkage between one paragraph and the subsequent paragraph is made.

14.4.4 Choice of Words

In addition to preparing a proper framework for the thesis, the paragraphs and sentences, it is vital that the researcher chooses the best possible words. A researcher gets a good vocabulary mainly from wide general reading, but other tools are available for those who need to improve this capability in a limited time. A good word processor will have a rather extensive thesaurus, and regular use of this will gradually improve the researcher's word choice. Some writing manuals have alphabetical lists of misused words which the researcher can always refer to. Many handbooks are available to guide researchers in technical report preparation. Reference to such books will be useful for researchers in this area.

While writing a thesis, the researcher should avoid use of vague words, unfamiliar words, needlessly complicated words, overworked words and phrases. At the time of proof-reading the researcher should look for unnecessary words and delete them. A word becomes vague when the researcher uses it in place of a correct specific word that is available. For instance, instead of using a specific word like hourly, daily or weekly where it is warranted, if the writer uses the word "periodically", then it becomes vague. Therefore, the word selection and context is very essential.

The researcher should not use words unnecessarily. For example a sentence is written as, "Through the use of finite element analysis, we are able to effect a solution of the problem". Unnecessarily many words have been used in this. A better way to write this sentence is, "we solved the problem using finite element analysis". Similarly if a word is very frequently used, it may lead to monotony in reading, therefore, a researcher should use proper alliteration of words. A thesaurus will be of much use in this.

Misuse of words have to be avoided to the extent possible. For example absolute words such as "perfect", "essential", "unique" and "certain" do not normally take modifiers. It is better to avoid phrases such as "very perfect", "most essential", "less unique", etc.

14.5 PRESENTATION OF GRAPHS, FIGURES AND TABLES

14.5.1 Display of Graphs

It may be necessary for the researcher to include graphs in the thesis, to improve the understanding. Reference material relating to graphs can be classified into two main types. One type concentrates on the principles guiding the direct visual communication between the graph and the reader. The other deals with methods used to analyse the meaning that may be hidden to the casual eye. The presentation graphs are used to display data for many purposes.

Nowadays computer-based software are extensively used for preparing graphs of various kinds; therefore the researcher should select the proper software for this. Scale lines (also called axes) form the horizontal and vertical borders of a rectangle called the data region. Some people instead of setting the two scale lines rectangularly (open type) enclose the data region in a box (closed type). Though it improves the elegance, it depends on the location of the graph in the page of the report. It is one's choice whether to have it or not. Along the scale lines tick marks may be placed to locate numerical values. The data region should be kept neat as far as possible.

Some researchers use grid lines while others do not. But if grid lines are felt necessary for some specific reason, they should be made light related to the bolder data curve. A key or legend has to be placed outside the data region to indicate some feature of data within the data region. Different symbols are used for plotting when several curves are plotted in the data region and this should also be included in the legend. Superimposition of curves can be made if it is necessary. But, if it is going to cluster, then they could be provided in several bands locating the curves side by side. When they are so located, the data labels are placed inside the data region in such a way that they do not interfere with the curves or other features.

Depending upon the need, markers which are brief notes usually outside the data region with arrows pointing to specific locations on the scale line. They are used to highlight certain aspects. The graph shown in Figure 14.1 indicates the effect of forging ratio on reduction of area of a rod both longitudinally and laterally in percentge. The marker details are also shown.

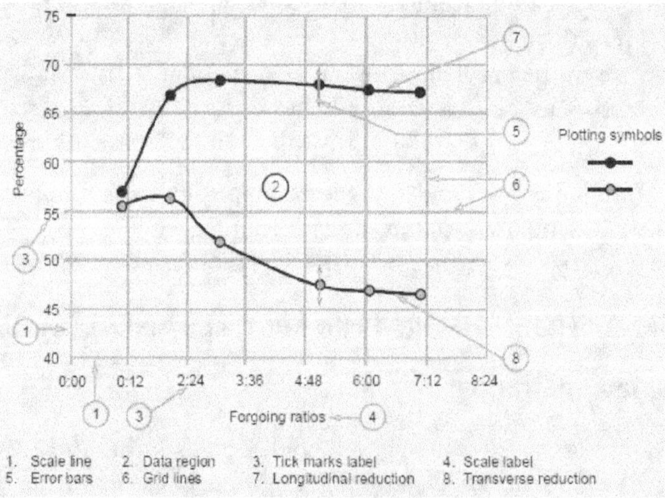

Figure 14.1   Graphic details

Error bars represent the 95% confidence limits when the data is subjected to statistical analysis. This provides both mean expected value and the measure of uncertainty. The researcher may want to include in the graphs some indication of uncertainty in the plotted points. Error bars are the most common means of doing this and methods for superimposing them on the basic data plots are provided in many commercial software. When error bar option is invoked, proper explanation of the displayed error bars shall be made in the report.

Many software packages provide three-dimensional graphical displays, which may look impressive. But the researcher has to be careful to ensure that the beauty of the display should not overshadow its real utility in communicating information. Generally two-dimensional graphs even though multiple in number provide clearer and more detailed information. Therefore, the researcher has to be careful in choosing a particular display depending upon the context and clarity required.

Various types of bar charts such as clustered bar, stacked bar and pie charts are available for the researcher to use. Instead of providing simply the bar or pie chart, if more details are needed, the researcher can add labels with exact quantity to each bar. One defect of pie chart is the poor angle-measuring capabilities of the human eye and mind; therefore, this can be easily overcome by simply labelling the pieces of pie with their percentage contributions.

14.5.2 Presentation of Figures

The term figures refers to line drawings, computer solid modelling, photographs and text charts. It is better to use line drawings because they are by far the most common in reports and papers. Line drawings can be classified into pictorial drawings, schematic diagrams and block diagrams. Pictorial drawings can be two-dimensional or three-dimensional and are simplified pictures of some real object(s). Schematic diagrams are more abstract than pictorial drawings, showing symbolic elements and their interconnection to make clear the configuration and/or operation of a system.

When making pictorial drawings, one should give enough thought to the process and his/her illustration should show only the needed items and not more. The advantage of a drawing compared to a photograph is that the researcher can select the required parts of the photographic view leaving the rest. A reasonably neat diagram can be drawn by hand faster than with graphics package and the researcher can scan it for an electronic version. Either of them is usually satisfactory.

Usually a one-bit (i.e., black and white), moderate resolution scan of hand-drawn sketch is bigger than a line drawing generated on a graphics package, but not huge. While considering the size of files, one should note that photographs may look pretty but take lot of memory. The major advantage is that a researcher can visualize all components, their interaction and then draw the schematic diagram. Hence the numerically small information content of a line drawing may provide more useful information than that of a photograph.

Despite line drawings being more informative than photographs, in certain cases photographs are desirable or essential. Surface details of metallurgical samples of failed specimens may be vital for proper interpretation. Many flow visualization methods use photographic records for data presentation and analysis. Therefore the researcher has to use line drawings, pictorial drawings, photographs, etc. optimally depending upon the situation.

Another important point for the information of researchers regarding figures and photographs is how to save them in the computer. In the digital version of the thesis, the researcher need not save ordinary photographs or other illustrations as bit maps, because they need more memory besides being slower in transfer. Many graphics packages allow saving in compressed form as .jpg (for photos) or .gif (for diagrams) files. Further the researcher can save space or speed up things by reducing the number of colours. In vector graphics (as used for drawings) compression is usually not necessary.

14.5.3   Display of Tables

The researcher can design tables with less difficulty compared to drawings and diagrams. Ruled tables are most common, but one can also use text tables and bulleted lists. A text table is usually short and is not numbered, titled or ruled into rows and columns; it is considered part of th text as shown below for an electrical strain gauge.

Strain gauge type	Sensitivity	Response to dynamic strain
Piezoelectric	High	Yes

Bulleted lists generally have one column and are not usually numbered. For instance,

The types of accelerometers are,

- Strain gauge type
- LVDT accelerometer
- Displacement type
- Piezoelectric
- Potentiometric

Ruled tables are major tables which are titled, numbered and ruled into rows and columns. Table 14.1(a) gives one arrangement of effect of cold drawing on modulus of elasticity in GPa for different types of steel with variation in temperature in K.

Table 14.1(a)   Results of three alloying treatments

Temp. (K)	Carbon steel (GPa)	Stainless steel (GPa)	Titanium alloy (GPa)
470	187	176	98
700	155	160	73
810	134	156	70

The above table can be rearranged as given in Table 14.1(b) given below:

Table 4.1(b)   Effect of temperature on modulus of elasticity

Details of steels	Temperature		
	470 K	700 K	810 K
Carbon steel (GPA)	187	155	134
Stainless steel (GPa)	176	160	156
Titanium alloy (GPa)	98	73	70

It may be noted from the Table 14.1(b) that it is easier to compare the modulus of easticity of different steels at various temperatures.

15
Structure of Thesis Report

15.1 SUGGESTED FRAMEWORK OF THE REPORT

The previous chapter discussed the plan for preparing a research report and the mechanics of writing. Issues such as style, presentation of figure and tables were dealt with. In this chapter we present a suggested framework for writing a thesis or dissertation keeping in mind the framework generally adopted by many universities for submission of thesis by research scholars. The proposed structure covers various sections to be included, besides reference listing, footnotes and glossary. The structure suggested is only indicative and the sections of the report may vary depending upon the nature of the research study. For instance, some universities may require the research scholar to include an abstract at the beginning of the report, while some other institutions prescribe to include a summary at the end of the main body of the report.

As stated in the section on thesis outline, the thesis structure refers to how the various parts of the thesis should be arranged. It is a standalone document. There is no single framework, which is suitable to all types of theses. Depending on the nature of study, the list of contents and chapter headings will vary. The researchers should consider their plan of the chapters and decide what is best to report in their work. They should make detailed lists of points that are relevant to headings and sub-headings or even the paragraphs of their theses. The logic of the presentation within each chapter should also be taken into consideration. However, a general structure of the thesis may not differ much. A suggested structure of a thesis is given in the following sections.

- Preliminary pages
    - Declaration/Bonafide certificate
    - Title page
    - Abstract
    - Acknowledgements
    - Table of contents

- List of figures and tables
- Definition of symbols and abbreviations used
- Main Body of Thesis
  - Introduction
  - Problem statement
  - Literature review
  - Core chapters
  - Findings and discussions
  - Conclusions
  - Suggestion for future work
  - Appendices
- References

The above layout of thesis is only an indicative one. Some universities specify that a summary is to be included at the end or at the beginning in lieu of an abstract. References are included after appendix by some or vice-versa by some others. However summary and footnotes are covered in this chapter.

15.2 PRELIMINARY PAGES

15.2.1 Declaration/Bonafide Certificate

The researcher has to make a declaration using the phrases recommended by the university/institution. Many universities have standardized the format for declaration. An example format runs like this,

"I hereby declare that this submission is my own work and that to the best of my knowledge and belief, it contains no material previously published or written by another person nor material, which to a substantial extent has been accepted for the award of any other degree or diploma of any university or institute of higher learning, except where due acknowledgement has been made in the text."

Signature/Name/Date should be appended below the certificate. It has to be a bonafide certificate by the supervisor.

15.2.2 Title Page

This page will contain the title of the thesis and author's name with an indication that it is being submitted for the degree of Ph.D./M.S in the concerned faculty/University and the date of submission. Again the format is prescribed by the university in many instances.

15.2.3 Abstract

Abstract is an important part of the thesis. This is the part that many would read and would find a place in many future journals and in Dissertation Abstracts. Therefore, this should be drafted with utmost care. It is better to write the abstract after the finalization of the draft thesis and it may require several revisions. In fact it should be a distillation of the thesis; a concise description of the problem(s) addressed, the method adopted to solve them, results and conclusions. An abstract must be self-contained and no references are included. In case a reference is necessary, its details should be included in the text of the abstract. It should be narrative and the number of words are not to exceed 600 or the limit prescribed by the university. Though it appears at the beginning, it is not an introduction and it is a resume of a thesis.

15.2.4 Acknowledgements

It is conventional to express gratitude to all those individuals or institutions that helped the researcher in the process (advice, scientific/technical/financial support, etc.) If the research work is a collaborative endeavour, the researcher should clearly indicate the contributions of individuals involved. A page may be devoted for this.

15.2.5 Table of Contents

The table of contents should contain chapter headings, sub-headings and their respective page number in a chronological order. It is conventional to use Arabic numerals in numbering from the introduction page till the end of the thesis and Roman numerals for the pages before the introduction page. As the thesis will be used as a reference, the table of contents helps to locate matters easily.

15.2.6 List of Figures and Tables

These are provided to help the reader to locate these features when they are referred to in a text location remote from the feature location. That is, if Figure 12.3 appears on page 223 but is later referred to on page 270, the reader should be able to locate it quickly. The list should use exactly the same captions as they appear below the figures. The same is applicable to tables also.

15.2.7 Definition of Symbols and Abbreviations

Definition of symbols and abbreviations used in the thesis provides clarity to the reader. They include both word description and the proper units. However, inclusion of this section is optional and may depend on the necessity for providing such a list. Especially in research studies involving science and engineering, a lot of mathematical symbols and notations are used. Hence, it is imperative that they have to be properly defined by the research scholar for easy understanding.

15.3 MAIN BODY OF THESIS

15.3.1 Introduction

This should cover the significance of the research topic and its place in the overall scheme of things in the vast discipline of the researcher. The relevance of the study and the problem statement should be couched in a fascinating style and in a language that a non-expert in the particular area of the study can easily follow. The introduction should arouse curiosity and interest in the mind of the reader to go through the thesis. In fact, the researcher should not over-estimate the reader's familiarity with the topic.

The introduction is primarily to introduce the research project to the reader. Hence it should contain a clear statement of the objectives of research which provides background to the problem considered. A brief summary of other relevant recent research works may also be stated so that the present study could be seen in that context. Hypothesis, if any, and methodology adopted may be very briefly stated. Observations and analyses made should also form part of introduction. In addition the limitations if any are narrated.

The introduction should be made interesting so that it enthuses the reader to evince interest in studying the thesis. This section might go through several drafts to improve its readability, at the same time keeping it short. If necessary it can be revised after the thesis is written.

15.3.2 Problem Statement

It is imperative that the problem taken up for research study has to be explicitly stated in the thesis. In some theses, it is found that statement of the problem is covered under the introduction. But, however it is always better to devote a section for the statement of the problem. The problem definition lays down the objective, scope and boundary, etc. There is an elaborate discussion of how to define the problem in Section 2.3 of Chapter 2.

15.3.3 Literature Review

The literature review is an essential part of research work. It should cover the genesis of the problem the researcher has taken up for study, and the state-of-the-art of knowledge already available about the problem. The literature review will have a fairly extensive list of significant papers or other works in the area along with the summary of their relevant findings. Further the researcher should link the background information obtained from literature reviewed with the proposed research problem. It was explained in detail in earlier chapter as to how a researcher should go about doing a literature survey. What is required is maintenance of proper record either in the form of a spreadsheet file or a word processor file by the researcher since the work is done over a period of 2 to 3 years.

Original research papers published in journals of repute provide an ideal source to reflect the state-of-the art of knowledge. Some authors feel that an order of 100 documents (papers) will be reasonable. But, however, it is a matter of judgment by the researcher as to how many papers should be scanned through and how relevant they are before they include them. Since the researcher is an expert on the (narrow) topic of his/her thesis, the literature review will demonstrate his/her expertise. A word of caution is that the researcher should not omit any relevant paper from which ideas have been extensively borrowed.

15.3.4 Core Chapters

The structure of the core or main body varies widely with the discipline and the nature of the topics. It may not be a single chapter in which all contents are put in by the researcher. For convenience and easy readability, the various chapters or sections under the core category have to be evolved. One cannot put forward a pattern to be followed universally.

The researcher should avoid the tendency to mechanically compile the papers, which he/she had published in journals based on certain aspects of his/her research work. There are many disadvantages in this approach. The journal articles cannot substitute the independent writing of a thesis by the researcher. The papers would have suffered the constraints of length and detail to comply with the norms of the concerned journals. Writing a thesis independently offers much more freedom in terms of length, breadth and depth of the treatment. As the researcher is the best judge, he/she alone can make the product self-supporting. It can stand out in quality without pressure from external constraints.

In certain theses it may be imperative for the researcher to establish some theory to justify the experimental techniques adopted. Thereafter, the researcher explains several different problems or different stages of the problem and finally presents a model or a new theory based on the work. For such cases, the core chapter headings could be theory, materials and methods, problem-1, problem-2, etc., and proposed theory/model. For others it might be appropriate to discuss different techniques in different chapters.

Materials and Methods vary enormously from thesis to thesis and may be absent in theoretical theses. It should be possible for another research scholar to replicate exactly what the present researcher has done by following the description available in the thesis. Therefore, it is imperative that the thesis covers a detailed description of the experimental set-up, arrangement, equipment used and their limitations if any and the procedure adopted in conducting the experiment. Similarly in the case of simulation study the core chapters should have description of the model built, assumptions made, model manipulation and extracting the responses from the model. Such details are necessary as another research scholar may do a similar study or experiment either in the same university or in a new set-up elsewhere.

When a researcher reports a theoretical work that is not original, he/she should include sufficient material to enable the reader to understand the arguments used and their physical basis. Sometimes, the researcher may present the theory from the beginning but should not reproduce pages of material that the reader could find in standard textbooks. Theory, which is not related to the researcher's work should not be included. The researcher should concentrate more on the physical arguments such as equations.

While reporting theoretical work the researcher can include more detail. However, lengthy derivation can be moved to appendices if it is likely to affect readability. Order and style of presentation is very important.

In sum, the core chapters should contain detailed description of the theories and hypotheses involved, different problems and the methodology adopted in resolving them, the materials and methods used, the experimental investigations carried out, the diverse techniques employed in the research study, significant revelations arising out of the work and arguments that led the researcher to the conclusions. Arguments may be conveniently presented as a series of numbered or bulleted points rather than as one chunk in a crowded paragraph.

15.3.5 Findings and Discussions

Discussion and findings are an important component of a dissertation as it brings out the significance, relationship, interpretation and generalization on the basis of experimental or study results. In fact, it is a challenging task for the research scholar. This requires the researcher to have not only familiarity with pertinent and up-to-date literature but also proper insight and knowledge of relevant concepts. Therefore it is advisable to write this section after completing other chapters since the main purpose is to discuss how the objectives of the study or experiment are achieved.

The findings and discussions are often combined in theses as otherwise the findings have to be repeated from core chapters for the purpose of discussions. While the major findings have already been presented earlier, this section provides a more complete treatment of all the findings (results) and their significance. The discussion should centre on the findings only and not on general discussion of the subject area. The nature of the discussion depends upon the specific research study. However, some general comments would be in order.

Findings are usually a combination of verbal statements, numerical values, tables and graphs. In some cases the researcher may give the same result both in tabular and graphical form. The graph helps one to discern the behaviour and trends quickly, whereas the table is better for giving accurate numerical values. If there is a theoretical prediction for a result which was measured, the researcher should compare theory and experiment and discuss their agreement or disagreement.

Discussion and findings need not follow the sequence of results obtained, rather it should follow a logical sequence relating to the objectives of the study. Some of the aspects that one should keep in mind while writing this section include,

- How the results contribute to the existing state-of-the-art of knowledge as evident from literature.

- Consistency with current theories and concepts on the issues investigated in the study.

- Whether the results give any new insights that lead to any new theories.

- Whether the results indicate any unexpected problem which needs to be explained.

- Discussion on results should be in terms of the original hypothesis the researcher proposed in the study.

- Instead of mere recapitulation of results, the discussion should cover principles, relationship and generalization based on results.

- Whether the findings have any theoretical or practical implications which warrants further research.

- Whether any limitations of the experimental set-up, or other aspects, affected the results.

- If there are any original contributions of the researcher by way of design and fabrication or new experimental set-up, this has to be justified for its validity and suitability for the study.

Apart from comparing theory and experiment, the researcher must compare and relate his/her findings to those of other researchers in the field, if such results are available. If there are significant discrepancies, the researcher should explain adducible reasons from his/her study. The practical importance of the findings of a research study in terms of converting it into a product, performance improvement, cost reduction or increase in productivity should also be discussed.

Discussion on error analysis should also be incorporated. The researcher uses error bar on the data unless the errors are very small. For single measurements, the bars should be on the basis of best estimate by the researcher in each coordinate. For multiple measurements, these should include standard error in the data.

15.3.6 Conclusions

In a research thesis, Conclusion refers to the body of logical inferences deduced based on the findings. Here, the researcher draws together all the findings and explains what conclusions have been reached based on the results and findings. These conclusions must be justified in terms of logical arguments, correlation with theory and practice, and the researcher's own judgments. It is imperative to make clear which statements are made based on facts and which are opinions. All conclusions must be supported by evidence and nothing new should be introduced at this stage. In the abstract, the researcher should include conclusions in a very brief form, because it must also include other material. A summary of conclusions is usually longer and the researcher has more space to be more explicit and more careful with qualifications in this section. It is better to put the conclusions in point form.

The Conclusions section is in fact a summary of significant findings from the result which leads to the conclusion. There may be several conclusions arising out of a research study. Each conclusion has to be supported by evidence based on factual findings and interpretation. A researcher should not over-generalize the findings and the conclusion from that. For instance, the mean of the first group is found to be significantly larger than the mean of the second group based on statistical test of significance results. On the basis of this test, the researcher can conclude that the original hypothesis is supported leading to the inference that the method employed by the first group is more effective than that of the second group. In this case no generalization has been made.

Over-generalization, on the other hand, refers to the statement of conclusions that are not warranted by the results and findings. For example, a particular variety of steel subject to cryogenic treatment process-A found to have significantly higher hardness than another cryogenic treatment process-B based on statistical significant test. It would be over-generalization to conclude that the heat treatment process of A is the best method for all varieties of steel, because only one variety of steel is subjected to test.

15.3.7 Suggestions for Future Work

In respect of scientific investigations, often more questions are raised than answers being produced. Of course, it is true that any topic can be explored endlessly as the human capacity is unlimited. It is conventional that a researcher suggests any interesting scope for further work by future researchers. The researcher should also indicate the practical implication of the work and limitations faced by him/her so that future research workers can try to address some of them.

In the scope for future work, the researcher has more freedom to express opinions that are not necessarily the direct outcome of data analysis and results. He/she can discuss any possible revision or additions to the theory and encourage studies designed to test the hypothesis suggested by the results. The researcher may also indicate the scope of the study replicable in other settings and other disciplines so as to improve the generality of his/her findings. Further the research scholar is free to propose the next step studies designed to investigate another dimension of the problem he/she investigated.

Usually this section is reasonably short—a few pages. As with introduction, it is a good idea to ask someone else who is not a specialist to read this section and offer their comments before finalization.

15.4   SUMMARY

As mentioned previously an institution may prescribe to include a Summary of the research work carried out by the researcher instead of an abstract of the thesis. Usually an abstract precedes the main body of the report and is located in the beginning of the thesis while the summary follows the section on discussion and findings and located at the end of the main body of the report. Sometimes the summary is also titled as summary and conclusions, combining both. Though abstract and summary appear identical, there are differences. A researcher, therefore, has to understand the distinction between an abstract and a summary of the thesis.

The main focus of an abstract is more on the problem and solution, so that it enables the researcher to identify the basic content of the report. It is an abstraction of the report in a nutshell and the size is always small. On the other hand, a Summary of thesis is more elaborate and covers almost all the chapters or sections of the thesis. The Summary gives the gist of the entire thesis including the problem addressed, methodology adopted for the experimental study, important observations, findings and conclusions. It should stand on its own so that it makes sense to the reader. In fact, by going through the summary, one should be able to get the full substance of the report.

Summary section normally does not have sub-headings or sub-divisions like a regular chapter of the thesis. Instead, it is organized into several paragraphs, which may be numbered. The size of the summary section also is more compared to an abstract and it may be about 2000 to 2500 words. Each paragraph has to be written in such a way that it brings out the essence of the problem taken up for study, key observations, and findings which led to the conclusions.

15.5   APPENDICES

Appendices are used sometimes in dissertation reports depending on the need. They provide supplementary information and data pertinent to a study, which either are not important enough to be included in the main body of the report or are too lengthy. Thus, appendices are useful for documenting information of lesser importance. By removing them from the main text, not only does the complexity of the flow of presentation get reduced but also the readability gets enhanced. For instance, if a detailed derivation of a lengthy mathematical model is included in the main body of the report, while the result of the model is of main interest to the researcher, definitely the readability gets disturbed. Such items have to be included as an appendix only.

In a dissertation there are several items, which a researcher can include under the appendix. The following items are typical examples for inclusion under appendix.

- Original computer programmes developed by the researcher
- Data files that are too large to be included in the Results section of the main body
- Pictures or diagrams of results which are not important enough for inclusion in the main text
- Long tables occupying more pages in the main body
- References the researcher had used but not cited in the main body of the text
- Detailed protocols of procedures for better appreciation of the reader
- Questionnaire used for the survey as a part of study process.

Appendices should be numbered using Arabic numerals such as Appendix-1, Appendix-2, and so on. The researcher can refer an appendix by its number in the main body of the text, whenever the matter presented is relevant for discussion. The location of appendixes in the dissertation report should be logically before the Reference section. This is because the Appendix may also have one or more references cited in it.

15.6 REFERENCES

15.6.1 References and Bibliography

A researcher while carrying out a study goes through several books, published articles, unpublished theses, conference proceedings, handbooks, web pages, private communication, etc. He/she may refer many points culled out from them in the thesis to support his/her view or to build further on them. All references which have been cited in the main body of the report are placed under list of references in a manner prescribed by universities or institutions to whom the thesis is submitted. Such listing enables the researcher not only to cite relevant material at appropriate places of the thesis but also to indicate the origin of the material by including under references, giving credit to the concerned author(s).

At times the term bibliography is also synonymously used with references. There are subtle differences between them. The bibliography lists all works which the researcher has read and to which he/she is indebted for ideas or information in general terms. On the other hand the purpose of the list of references is to point out the specific location of an idea or a piece of information in the original source. The research scholar prepares the bibliography before writing the report to remember the work he/she intends to consult whereas the list of references is more conveniently prepared while the report is being written.

15.6.2 Format of Listing of References

Listing of references has to follow the referencing system adopted as discussed in the chapter on writing journal papers and synopsis. Whatever we discussed there will be equally applicable for the thesis, but on a larger scale, since it covers the entire report. The listing format for the thesis is mostly governed by the guidelines provided by the university. Several professional bodies such as Institution of Electrical and Electronics Engineers (IEEE), editors belonging to disciplines like Biology, Psychology, etc., prescribed their formats. It appears that there is no single format which is universally accepted by all for reference listing.

A format followed by many scientific and engineering communities and adopted by many universities for dissertation is given below through examples.

1. *Books referred*

    i. Dym, C.L., "Engineering Design: A synthesis of views," New York: Cambridge University Press, 1994.

    ii. Kahnemann, D., Slovic, D.P. and Tversky, A., Judgement under uncertainty: Heuristics and Biases, Cambridge, England: Cambridge University Press, 1982.

2. *Journal Publications*

    i. Clark, C., "The Utility of Statistics of Random Numbers", Journal of Operations Research, Vol.8, No.2, March-April, 1962, pp. 20-25.

    ii. Luxhol, J.T., and Hansen, P.H.K.,"Engineering Curriculum Reform at Aalborg", Journal of Engineering Education, Vol. 85, No.3, 1996, pp. 83-88.

3. *Paper in Conference proceedings*

    i. Diananda, P.H.,"Some Probability Limit Theorems with Statistical Applications", Proceedings of Cambridge Philosophical Society, Vol. 49, 1953, pp. 239-46.

    ii. Hatchuel, A., and Weil, B., "A New Approach of Innovative Design: An Introduction to C.K.Theory," Proceedings, International Conference on Engineering Design, Stockholm, Sweden, 2003.

4. *Patents*

    i. Gustafsson, J.K., "Analog-digital converter for a resistance bridge", Patent U.S.3960010, June 1, 1976.

5. *Websites*

    A website may disappear or it may have been updated or changed completely. So reference to the website is usually less satisfactory. Nevertheless, there are some very useful authoritative sources. So if the rules of the university permit, a researcher may include it in the references cited.

    i. Data Sheet, PIC 16F87XA,"28/40/44 Pin Enhanced Flash Microcontrollers," Microchip Technology inc., 2003, http:// ww1.microchip.com/downloads/en/Device Doc/395826. pdf

       ii.   Mc Laughlin, Michael P, "A Tutorial on Mathematical Modeling". 1999 http://www,causascientia.org/math_stat/ Tutorial. Pdf

6.   *Handbooks referred*

       i.   Beitz, W., and Kitter, K.H.,(eds), "Handbook of Mechanical Engineering", Springer-Verlag, London, 1994.

       ii.   Lingaiah, K., "Machine Design Handbook", McGraw-Hill, New York, 1994.

7.   Private Communication

       i.   Doe, J.J., XYZ Company, Atlanta, PA, unpublished research, 1981.

       ii.   Catz, J., Personal communication, Department of Mechanical Engineering, University of Miami, Florida, 1987.

15.6.3 Footnotes

Footnotes are used specifically to validate a point, a statement to supplement or amplify materials included in the main body. It provides the reader with sufficient additional information to enable them to consult sources independently. Neither in the abstract nor in the main body this will appear and it appears only at the bottom of the page and/or beneath the table. The footnote below the table is normally used to indicate the source of data, information pertaining to the entire table or to include necessary comments on specific items. Also, when the number of references made is only a few, instead of listing them under reference section, they can be included as footnotes at the bottom of the respective page itself. If there are many footnotes they can be listed under endnotes at the end of the report.

All the footnotes are marked by using appropriate superscripts in the body of the text where a particular reference is made. While placing the footnotes at the bottom of a page, they should be separated from the text by a fifteen space line drawn from the starting of the margin below the last line of the text. If a footnote occupies more than one line, they should be single-spaced and the first line is indented. Many conventions are adopted in using footnotes.

Research studies of science and engineering have many tables based on data gathered and collated, and are used to draw inferences. Sometimes data from secondary sources are also utilized. Therefore, use of footnotes below the table to indicate the original sources of data will enhance the validity of the report. However, for referencing whether footnote could be used or not depends on the prescriptions of the university. While using a footnote for explaining or supplementing a point, it should be concise and legible without sacrificing brevity. If footnotes are used in the thesis, consistency has to be maintained throughout the dissertation.

15.7   GLOSSARY

A scientific or engineering research study involves lot of technical jargon or special terms. In the case of applied research involving an industrial problem, certain terms peculiar to that domain will have to be used. Therefore, it is necessary to include a glossary wherein all the technical words and phrases are defined and explained. If it involves only a very few words, perhaps, they can be explained by way of footnotes at appropriate places of the main text itself.

Whether the researcher should include a glossary in the thesis depends upon who is going to read his/her report. If the field of expertise of readers is the same as that of the researcher there may not be any necessity for a glossary. But if audience is from other areas, it is advisable to provide a glossary. This is required from the thesis examiners point also. Even though a thesis evaluator is an expert in that subject, when applied research works are undertaken, jargon used pertaining to that industry has to be explained so that the examiner ould understand and appreciate better.

16
Evaluation of a thesis

16.1 CURRENT PRACTICE

An essential requirement for the award of Doctor of Philosophy (Ph.D.) degree to a researcher by a University is preparation and submission of a thesis or dissertation based on the research work carried out by him/her. This practice is prevalent in almost all recognized universities across the world, though wide variations are there in other requirements such as number of courses of study to be undertaken, number of credits to be earned and duration, etc. A majority of universities do insist on certain minimum number of courses to undergo (a decision by the supervising committee) and minimum credits to be earned. Thus the learning objectives and the learning process for doctoral study are not articulated in the conventional ways.

By acquiring a doctoral qualification one reaches the zenith of formal education. This enables him/her to acquire international recognition besides elite status in the academia. Therefore, a research scholar is expected to provide an original and significant contribution based on his/her research work in the form of a thesis. Neither are there universal guidelines to recognize as to what constitutes appropriate research nor are there benchmarks for research outcome at the doctoral level. Similarly prescribing standards for the thesis and its evaluation process have been a problem for the academic community. In this background, the standard of evaluation solely depends on the ability and knowledge of the examiners who adjudicate theses.

Inasmuch as there is no curriculum for Ph.D., it is the thesis examiners who really set the yardstick of what is acceptable as thesis or dissertation, thereby setting the standards for conferment of Ph.D. degree to a scholar. While universities do prescribe format and guidelines for preparation of thesis, to a large extent the examiners' observations and comments influence the style and contents to be included in the thesis report. The examiners' report is specific to a particular thesis and after decision it is confined to the university and not made public. Unless the university or some other common academic body conducts an in-depth study of such reports and brings out the essential features in a general format, the research fraternity is denied the benefit of such feedback.

Universities in India as followed elsewhere refer the thesis to two or three examiners for independent adjudication. Some universities refer to two and others to three adjudicators, nominating at least one from a foreign country and the rest within India. By getting independent assessment from each examiner without prior consultation between them, the university has the benefit of multiple independent assessments for a final decision regarding the thesis. Based on these independent assessments, the university's appropriate body is free to decide whether to accept or reject the report and to confer the doctoral degree to a researcher subject to the compliance of other formalities.

From the evaluation format of many universities, it is noticed that the final recommendation of the examiners of thesis can be grouped under five categories as given below:

i.   Thesis may be accepted as submitted by the researcher and the examiners accord suitable commendation such as highly commendable, commendable, etc.

ii. Thesis may be accepted after minor corrections suggested by the examiner are incorporated.

iii. Thesis may be accepted subject to the condition that major corrections pointed out are incorporated.

iv. Thesis has to be revised incorporating the suggestions, modifications, and answers to queries and resubmitted to the same examiner for further evaluation.

v.   Thesis is not of acceptable standard, having no opportunity for revision.

It is evident from the above range of recommendation of examiners, that a researcher, has adequate opportunity to correct and revise the work taking into account the examiners comment. This is one of the prime reasons why the number of failures in Ph.D. programme is much less.

In addition to evaluation of thesis by two or three examiners independently, a researcher is required to appear for an oral defence examination before a panel of examiners constituted by the university which invariably include at least one examiner who had evaluated the thesis besides another examiner nominated afresh by the university and the supervisor. Before the researcher appears for oral defence examination, all the suggestions/corrections indicated vide item (ii) and (iii) supra have to be incorporated in the thesis. Regarding item (iv), the researcher can face the oral examination only after getting positive report of the examiner on re-submission of the thesis. Apart from the panel of examiners, the audience is also free to raise queries and the researcher is bound to clarify and defend his/her work.

16.2 ENSURING QUALITY IN RESEARCH

Research is an intellectual activity of an individual. A doctoral scholar's dissertation reflects his/her maturity in the overall process, study planning, data collection, analysis and inferences and written communication, all playing significant role equally. In fact, it emphasizes the researcher's ability to draw different parts together. Therefore, the primary responsibility for the production of quality thesis rests with the researcher and this ability depends on his/her intellectual acumen, commitment, tenacity and willingness to do hard work.

Though the primary responsibility for quality research work rests with the researcher, built-in checks are there in the doctoral programme of every university. For each researcher, a research supervisor is assigned; besides a supervisory committee comprising three or four experts is formed by the university. While the supervisor is supposed to oversee the work of the researcher on a day-to-day basis, the supervisory committee (Doctoral Committee) meets periodically and/or at appropriate milestones such as confirmation of registration, approval of synopsis to scrutinize the work of the scholar. Thus, quality check is ensured at three levels, viz., at the scholar level self-evaluation, at the supervisor level and at the supervisory committee level.

As the research work is carried out over a period of 3 to 4 years, the scholar has enough opportunities to present papers in conferences, both national and international, and also publish papers in peer-reviewed journals based on his/her research work. When a paper is sent for publication in a peer-reviewed journal, the review comments provide enough scope for improving the quality of work. Besides, the queries raised by the audience, discussions and interaction during conferences help the researcher with valuable insights, to improve his/her research content. Realizing the importance of publication, universities are making it mandatory to publish at least one paper arising out of research work in a refereed journal with impact factor.

Despite all these built-in checks and balances, one often hears that theses produced are lacking in quality. Not only the quality but also the number of research papers published in refereed journals by Indians are much less compared to those from developed countries like USA, UK, Japan, etc.

Not many studies have been reported taking into account the examiners evaluation report of theses from various universities, generalizing their observations and comments, and evolve suitable guidelines for the benefit of research scholars. In this context, it is worth referring to the study undertaken by Holbrook, A., Bourke, S., Lovat, T., and Dally, K., of University of New Castle, Australia, a national project sponsored by Australian Research Council, wherein they analysed reports of 803 examiners involving 301 theses of Australian Universities and came out with findings for the use of researchers.

To ensure quality in research, proper attention coupled with monitoring and periodical reviews has to be exercised right from the problem identification stage till the scholar submits the thesis. Lack of commitment and a complacent attitude are the prime factors both on the part of scholars and supervisory structure for the mediocre quality of research work. Another reason is that not many bright youngsters are willing to join research-oriented programmes, because of better prospects elsewhere. The government and national academic bodies have to devote attention to this. Moreover periodical feedback has to be provided by analysing reports of examiners, to provide general guidance to new entrants. Perhaps certain indicators may be developed involving different assessment traits of the thesis to ensure objective assessment by the examiners.

16.3 COMMENTS AND SUGGESTIONS

A research thesis is a standalone document pertaining to the problem investigated by a scholar. Nevertheless, the methodology adopted, data collection, analysis made and inferences drawn must be in accordance with generally accepted principles and practices that enable one to produce credible and verifiable results. A thesis should add value to the existing literature by expanding the frontiers of knowledge and should reflect organized thinking and insights. Rationale of new contributions, their significance and usefulness to the society must be covered. Though the comments of examiners are specific to a particular thesis, generalizing the comments of various theses will provide a useful feedback to scholars.

The following section explains how a thesis can be evaluated/examined taking into account the comments of examiners' reports of theses, information available from literature, personal discussions with senior professors who are involved in evaluation of theses for decades. They are detailed under the following headings: problem taken up, literature reviewed, methodology, analysis and interpretation, contributions made and thesis presentation. This can even serve as a checklist to refer while researchers prepare their theses.

i. *Problem considered* Problem formulation and statement in a thesis is the prime aspect from where other things follow. The key issues to examine include,

- Whether the problem identified is investigable through collection and analysis of data.
- Whether the problem objectives and scope are set forth properly.
- Whether the relevance of the problem in the current situation is established.
- Whether the significance of the problem involving practical or theoretical aspects is indicated.
- Whether a problem selected for study is justified on the basis of information such as citation from primary literature.
- Whether the problem statement identifies variables of interest to be investigated and the specific relationship between them.
- Whether all the variables of interest are defined properly.

If the problem perspective is not explained properly, it gives an impression to the examiner that the researcher has not comprehended the problem and is not very clear as to what he/she proposes to do. Only by a proper statement of the problem the main focus of the scholar will be evident to the examiner. Sometimes, the problem formulated by a scholar is an excellent one for research, but poor written communication indicates lack of grasp. Therefore, it is not enough if a problem is well-conceived but it has to be properly explained in the thesis.

ii. *Literature review*   Review of literature not only gives an overview but also demonstrates concisely the level of understanding of research related to the issues focused in the problem. Hence, an in-depth discussion of literature reviewed must be incorporated in the thesis. Major issues of relevance are,

- Whether the review of literature is comprehensive and relevant to the problem taken up for investigation.

- Whether most of the sources are from primary or a few are from primary and the rest from secondary sources. Expectation of examiners are that original papers should be consulted and that too recent ones.

- Whether the references are critically analysed, compared and contrasted with the results of various studies. Sometimes a researcher may include all references whether relevant or not. Such action may lead to comments like "... has failed to recognize useful and reject not useful research..."; "... has not engaged adequately with a body of relevant knowledge..." etc.

- Whether the review is organized in a logical fashion with least related references discussed first and the most related ones discussed last.

- Whether review made is concluded with a brief summary of literature relating their implication with the problem taken up.

- Whether the implications discussed above form an empirical or theoretical rationale for the hypothesis which he/she formulated.

Summarization and analysis of implications have to be accurate and should not result in misleading interpretations. The examiners with their expert knowledge in the area are able to see through this and may adversely comment if misleading interpretations are made. Such misleading interpretations by a scholar give the impression to the examiner that the researcher's comprehension is of mediocre quality.

iii. *Methodology adopted*   Here, the issues covered include formation of research hypothesis, methods employed for data collection, analysis made and results arrived at. Methods employed for the study may include experimental set-up designed or simulation model developed for manipulation and extraction of data. Significant aspects that a researcher should bear in mind include the following.

- Whether any specific testable hypothesis or any specific questions to be answered through the study are listed.
- Whether the methodology proposed to be adopted is quite appropriate vis-à-vis the approach to the research questions framed.
- Whether the methods envisaged are well-justified, sound and clearly described.
- In case the study contemplates any experimental set-up designed exclusively for the study, whether procedures involved for the development and validation are adequately justified.
- Whether reasons for the use of any specific equipment for taking measurement explained or not.
- How the models developed for research studies have been validated before using them for experimentation.
- Whether rationale of abstractions and assumptions made in model development and the methods adopted are adequately explained.
- Whether the procedures described are sufficient to permit replication by another researcher.
- Whether the experimental design adopted is appropriate to the problem taken up for study.
- Whether justification is given for use of statistical techniques used in the analysis of data and the appropriateness of the results achieved.
- Whether any pilot study is conducted to fix the factor levels described with its impact on subsequent results.
- Whether extraction of results and tabulation and analysis adequately described in the text of the report.
- In case there are any weaknesses or limitations in the method, they should be spelt out. It should not lead to identification by the examiner with an adverse comment.

Examiners may be unwilling to accept a thesis if scholars did not demonstrate adequate knowledge of their topic, and a thorough understanding of their methodology and limitation.

iv. *Analysis and interpretation*   Analysis and interpretation of results forms the crux of a research work. Here the researcher is able to link his/her results with the theoretical background knowledge justifying interpretation and conclusion made. Various aspects which a scholar should consider while doing analysis and interpretation are,

- Whether data analysis method employed is appropriate and fully justified. Inappropriate data analysis results in comments such as, "... offers no critical analysis of information presented"; "... understanding the methodology appears to be rudimentary ...".

- Whether each result discussed in terms of its agreement or disagreement with the previous results obtained by other researchers.

- Are the theoretical and practical implications of interpretations discussed sufficiently?

- Whether the analysis and interpretations made are in terms of original hypothesis formulated or questions raised for the study.

- Whether the interpretations are backed up by evidence or claims are overstated about the generality of interpretations/findings.

- Whether the scholar demonstrates his/her capability by proper analysis and interpretations making critical and original insights.

v. *Contributions* An important expectation of an examiner in a Ph.D. dissertation is the scholar's original contribution arising out of his/her study. The contribution includes originality, substance, impact and suggestions for future study. Though many scholars do not mention contribution and significance explicitly in their theses, the examiner looks for this in every dissertation. Important considerations a researcher should have are the following.

- Whether the scholar demonstrated his/her contribution in each chapter of the thesis by adopting a unique approach in presenting the material in a thought-provoking manner.

- Whether any rich insights have been made in his/her work which reflect scholarly and impressive contribution. Proper insight leads to major contributions to the discipline.

- Are conclusions made likely to lead to the advancement of knowledge in the field and that they are not misleading?

- Has the contribution of research resulted in practical outcomes?

- Whether the new design of an experimental set-up, is original in nature and justified from the literature reviewed.

- Any improvement to existing algorithm or process or design of new process or algorithm cited as original contribution, justified with supporting evidence.

- If the research work is in the cutting-edge area, whether the scholar covered its importance in the report as it involves challenging issues.

- Whether scope for extension of existing research to further research indicated. Any unanswered questions or limitation stated are likely to be the sources for other researchers to make further studies.

vi. *Presentation* Even if a thesis is of high quality, if the presentation is poor, the examiners will comment, "... the writing and presentation of thesis is substantially below the level ...". Therefore a scholar has to ensure a good presentation of research work. The important points to be kept in mind are as follows:

- Whether communication of facts are cogent and straightforward to make it easy to read.

- Whether the flow is logical and not leading to poor communication of ideas.

- Are all necessary steps in the argument supporting an essential point made crystal clear.

- Whether care has been taken to ensure proper grammar and flawless expression in the writing which makes the presentation better and error-free.

If a thesis is error-ridden the examiner may get irritated and evaluation gets affected even if the work is of commendable nature. Examiners tend to interpret the capacity of the scholar to disseminate information proficiently, logically and concisely in the dissertation as an indication of the scholar's thorough understanding of his/her subject.

A high-quality thesis can be characterized by the development of a clear conceptual framework; research questions and hypotheses which are clearly stated and carefully explained; a thorough literature review which synthesizes the relevant information and demonstrates a sure grasp of theoretical debates; a sound methodology which is adequately justified; systematic and logical analysis, and findings that are well-supported and linked to the theoretical analysis.

16.4 THESIS DEFENCE EXAMINATION

16.4.1 Preparation for Oral Presentation

After the thesis is evaluated by the examiners nominated by the university, the researcher has to appear before a viva-voce examination board constituted by the University to present his/her research work and defend it. This is an important stage in the Ph.D. research project. Mostly it is a open viva, where after examination by the examination board, the audience may also raise queries and the researcher is expected to answer their queries satisfactorily. The researcher should not consider this session as a confrontation by a group of experts against him/her. The researcher should take it just as an opportunity for presentation of ideas which he/she extensively nurtured during the last three to four years.

As a person who carried out the research work, the researcher is sure of all the diverse aspects of the topic, perhaps far better than the members of the examination board before him/her. The objective of the researcher is to convince the audience that the job has been done well, the methodology is sound and the findings are useful and original. In experimental topics, all the findings of the scholar should be based on repeated experiments, which should be replicable by anyone else.

It is better for the researcher to know in advance the time duration given for making the presentation so that one can plan and make use of the time allotted most effectively. If a researcher is asked to finish the presentation say, in the next two minutes, when he/she is only midway, the whole effort will end up badly bruised. This has to be avoided at all costs. Therefore a proper preparation is necessary. Normally a scholar is allowed to make a presentation for about 30 to 40 minutes before the panel of examiners and audience and thereafter, he/she has to answer queries raised by the panel members and then the audience.

If the researcher is allotted 30 to 40 minutes for oral presentation, he/she should make a plan as to how many slides could be presented using PowerPoint presentation. A minimum of 30 seconds should be left for each slide to enable the audience to read and see what is shown in the slide. Accordingly slides should be prepared to cover only the important aspects. The researcher should focus mainly on the problem taken up for study, how he/she approached the solution for the problem considered and major findings and conclusions. However, the researcher should have enough back-up slides, to answer any specific queries after the main presentation. After preparing the slides, it would be helpful if he/she rehearsed the entire presentation in the presence of the supervisor so that the latter's suggestions could be noted and appropriate changes to the final presentation made.

Discussion with friends and colleagues on various aspects relating to the thesis and gathering possible questions that they contemplate, would also be very useful to the researcher.

16.4.2 Oral Presentation and Defence

Viva-voce examination is an exercise on oral communication for the researcher. The oral communication has several special characteristics: quick feedback by question and dialogue; impact of personal enthusiasm; impact of visual aids; and the important influence of tone, emphasis, and gesture. A skilled speaker in close contact with an audience can communicate far more effectively than the cold, distant easily evaded written word. On the other hand, the organization and logic of presentation must be of a higher order for the oral than written communication.

Many opportunities for noise exist in oral communication. Proper preparation and delivery by the researcher coupled with good quality of visual aids can improve the efficiency of oral communication. Comments from others should be taken as valuable. So also, any new idea coming from the member of the examination panel should be welcomed.

A researcher should listen and comprehend the points made and/or queries raised before attempting to answer or clarify the points. Clarifications by the researcher should be clear and convincing. Some months would have elapsed between preparation and submission of thesis and holding of viva-voce examination. Therefore, before the viva-voce examination, the researcher should once again thoroughly go through the report so that he/she can answer any queries or clarify points raised. The researcher should never go into arguments.

While making a presentation, the scholar should maintain direct eye contact with the expert panel and sections of the audience. This gives confidence not only to the presenter but also demonstrates to the audience that the presenter is a confident person. Further, eye contacts provide an interaction between the audience and the scholar. The speaker should not stare at any particular individual in the audience for more than a few seconds. It should look natural and the presenter should shift focus on different sections. While making PowerPoint presentation reading of the slide has to be minimized. After all, the scholar is making a presentation of his/her research work and not reading it and hence should avoid creating this impression.

A researcher should approach the viva-voce examination with a feeling that the expert panel is there to offer help. This feeling would boost his/her confidence level. After all, the long effort of the researcher, it is reaching its concluding stage of success. In fact, the researcher is not alone—the supervisor also has a role in the entire process of reseach and thesis preparation.

17
Ethics in Research

17.1 IMPORTANCE OF ETHICS IN RESEARCH

Ethics can be defined as the system or code of morals of a particular person, religion, group or profession. Ethics is relative. It differs from one individual to another. Ethical behaviour is an individual matter and a person's ethics and morals are built into his/her personality right from childhood. However, the key forces that form the ethical behaviour of an individual includes his/her family, peer influence, experience logged by the individual, his/her morals and values practised and opportunities. Here, we look into ethical behaviour as individuals involved in scientific and technological research works.

Research ethics involves the application of fundamental ethical principles to a variety of topics related to scientific research. These include the design and implementation of research involving physical experimentation, human and animal experimentation, in-depth theoretical and other studies pertaining to various disciplines. It also includes various aspects of academic scandal including scientific misconduct such as fraud, fabrication of data and plagiarism, whistle blowing and regulation of research, etc. In the medical field research ethics is well-developed. Many professional societies like American Society of Mechanical Engineers (ASME) have also evolved ethical codes for their professionals to adopt.

Scientific research is built on a foundation of trust and hence scientists believe that the results reported by others are valid. The society also trusts that the results of research reflect an honest attempt by scientists to describe the phenomena and innovations accurately and without any bias. But this trust will endure only if the scientific community devotes itself to embodying and transmitting values associated with ethical conduct.

Many ethical issues have to be accorded serious consideration for research and they depend upon the nature of the research work. For instance, in the case of sociological research not only permission of those involved in the study need to be obtained but also their rights are to be protected and privacy and sensitivity are to be maintained. Research ethics in medical context is dominated by principalism. Nowadays many interdisciplinary native research projects are pursued, involving engineers, biologists or doctors where animals and humans are used for experimentation. Hence an engineer must be aware of ethical practices followed in medical and veterinary fields also. In research involving physical experimentation, sincerity in reporting, safety of the individuals involved, environmental considerations, etc. are important.

As science becomes increasingly intertwined with major social, philosophical, economic and political issues, scientists and engineers become more accountable to the larger society of which they are a part. Therefore individual scientists/engineers and their institutions should periodically reassess the values and professional practices that guide their research as well as their efforts to perform their work with high integrity.

17.2 INTEGRITY IN RESEARCH

17.2.1 Integrity of Researcher

Integrity in research is essential for maintaining scientific excellence and for ensuring public trust. It characterizes both researchers and the institutions in which they work. A committee set up by the National Academy of Sciences in USA observed, "For a scientist integrity embodies above all the individual's commitment to intellectual honesty and personal responsibility. It is an aspect of moral character and experience. For an institution, it is a commitment to creating an environment that promotes reasonable conduct by embracing standards of excellence, trustworthiness, and lawfulness..." Thus, integrity is an important trait both for individual researchers and for the institution in which they work.

For an individual researcher integrity embodies commitment to intellectual honesty and personal responsibility for one's actions. It also covers a range of practices that characterize the responsible conduct of research which include,

- Intellectual honesty in proposing, performing, and reporting research.
- Accuracy in representing contributions to research proposals and reports.
- Fairness in peer review
- Collegiality in scientific interactions including communication and sharing of resources.
- Transparency in conflict of interest or potential conflict of interest
- Ensuring protection and safety of human subjects in the conduct of research.
- Adherence to the mutual responsibilities between investigators.

Intellectual honesty requires that researchers present proposals and data faithfully and communicate their best understanding of their work in writing and verbally. They should advocate for their research conclusions in the face of collegial skepticism and the will to acknowledge errors. The researchers are not expected to report the work of others as if it was their own, which is plagiarism. Collegiality in scientific interactions requires that investigators report research findings to the scientific community in a full, open and timely fashion. At the same time, it should be recognized that the scientific community is highly competitive. The researcher who first reports new and important findings gets credit for the discovery. Therefore even in fierce competition, a researcher should not lose balance and be honest and keep his/her integrity both in work and reporting.

17.2.2 Integrity of Supervisor

The faculty advisor (supervisor) is the mentor for a young researcher. His/her responsibilities include a commitment to continuous education and guidance of trainees, appropriate delegation of responsibility, regular review and constructive appraisal, fair attribution of accomplishment and authorship, and career guidance. The supervisor should also help the research scholar in creating opportunities for employment and funding. For the new researcher the essential elements of ethics include respect for the faculty advisor, loyalty, strong commitment to science, dedication to the project, careful performance of experiments, precise and complete control of record keeping, accurate reporting of results and a commitment to oral and written presentations and publications.

17.2.3 Institutional Integrity

Institutions should always strive to create an environment that promotes responsible conduct by individual researchers that foster integrity. They must establish and continuously monitor structures, processes, policies and procedures that

- provide leadership in support of responsible conduct of research
- encourage respect for everyone involved in the research work
- promote productive interactions between research scholars and faculty advisors
- advocate adherence to rules regarding all aspects of the conduct of research
- manage individual and institutional conflict of interests
- take timely action against scientific misconduct
- educate the researchers pertaining to integrity in the conduct of research

Promotion of integrity through education is an important task of institutions. Integrity poses a special challenge because it requires a synthesis of ethics and science. When engineers and ethicists collaborate in the design and implementation of learning experiences, the research scholars come to appreciate the complexity of problems that arise in the practice of science. Research advisors play a central role in the education of their researchers in the responsible conduct of research, not only by what they teach, but also by their own conduct. The committee set by the National Academy of Sciences, USA, which probed into the question of research integrity emphasized the education of research students and post-doctoral fellows as they felt this was where the future lay. This is true for the Indian context as well.

It is the burden of research institutions to promote and monitor the responsible conduct of research. They should consistently provide the researchers with the resources they need to conduct research. These resources include leadership, training and education, policies and procedures as well as tools and support systems. Anyone needing assistance should have ready access to all knowledgeable sources in the institution.

17.3 SCIENTIFIC MISCONDUCT AND CONSEQUENCES

17.3.1 Scientific Misconduct

Another important aspect which is worthy of consideration is misconduct in scientific research. A scientific misconduct can be defined as "the intentional distortion of the research process by fabrication of data, text, hypothesis or methods from another researcher's manuscript form or publication; or distortion of research process in other ways". The consequences of scientific misconduct can be severe at a personal level for both perpetrators and any individual who exposes it. The main motivating factors for researchers to commit misconduct are,

- *Career pressure*   The reputation of a researcher lies on publication of high-profile scientific papers and getting research grants. This induces some persons to fabricate results.
- *Laziness*   Researchers introduce a fact that they believe is true without taking the trouble and difficulty of actually performing the experiments required.
- *Lure of money*   When grant is available for developing new technologies for which no known standards are available for comparison.
- *Ideological considerations*   Because of ideological commitments, a researcher may misrepresent facts intentionally.

Scientific misconduct is the violation of the standard codes of scholarly conduct and ethical behaviour in professional scientific research. The various forms of scientific misconduct is enumerated below.

- *Fabrication*   The publication of deliberately false or misleading research. This is subdivided into obfuscation (omission of critical data or results), fabrication (actual making up of research data), falsification (manipulation of research data and process), and bare assertion (making entirely unsubstantiated claims).
- *Plagiarism*   The act of taking credit or attempting to take credit for the work of another. A subset is citation plagiarism, i.e., willful or negligent failure to acknowledge appropriately other or previous discoverers. It is the most common type of misconduct.
- *Self-plagiarism*   Multiple publication of the same content with different titles and/or in different journals.
- *Violation*   It is the infringement of ethical standards regarding human and animal experiments.

- *Ghost writing*   The phenomenon where someone other than the named author(s) makes a major contribution.

- *Unjust credit*   This relates to conferring authorship on those who have not made substantial contributions to the research work. Often this is done by senior researchers who force their way on to the papers of inexperienced junior researchers as well as others that stake authorship in an effort to guarantee publication.

- *Misappropriation of data*   Literally stealing the work and results of others and publishing it so as to make it appear as if the author had performed all the work for which the data was obtained.

- *Suppression of data*   Some institutions consider hushing up data or findings as their publication will be adverse to their interests or the sponsors.

The authors and co-authors of scientific publications have a variety of responsibilities. Contravention of the rules of scientific authorship may lead to a charge of scientific misconduct. All authors are expected to keep all study data for later examination even after publication. The failure to keep data may be regarded as misconduct. Therefore all authors including co-authors are expected to make reasonable attempts to check findings submitted to academic journals for publication. Simultaneous submission of scientific findings to more than one journal or duplicate publication of findings is usually regarded as misconduct.

17.3.2 Consequences of Misconduct

In many countries there is no regulator to oversee the investigation of allegations of research misconduct. Universities are not taking much interest to investigate allegations or act on the findings of such investigations to vindicate the allegation.

The effect of scientific fraud depends on the severity of the fraud, the level of notice it receives and how long it goes undetected. For cases of fabricated evidence, the consequences can be wide-ranging with others working to confirm or refute the false finding or with research agendas being distorted to address the fraudulent evidence.

Regarding consequences of individuals indulged in misconduct it has reflection on the institutions that host or employ them and also on the participants in any peer review process that has allowed the publication of the questionable research. Several persons are involved in this. The persons who expose such cases can find themselves open to retaliation by a number of different means. These negative consequences for exposure of misconduct have driven the development of whistle blowers. In some countries like USA there are charters to protect those who raise concerns. A whistle blower is almost always alone in his fight and his/her career becomes completely dependent on the decision of alleged misconduct. Even some organizations feel shy of employing such persons on the pretext that they are trouble makers.

Nowadays there are several tools to aid in the detection of plagiarism and multiple publication, thanks to the development of Internet. A tool developed in 2006 by the University of Texas, called "Déjà vu" uses the text mining algorithm eTBLAST and identified many instances of duplicate publication in the biomedical area. Most tools that are used to detect falsified data include error analysis. Most measurements usually have a small amount of error. Suppose in a particular case, the amount of error normally be in such measurements does not appear, one can infer the data may have been forged. However, this alone is not sufficient for confirmation. Further investigatons have to be conducted.

18
Intellectual Property Rights

18.1 THE INTELLECTUAL PROPERTY

The word 'intellect' connotes the faculty of reasoning, knowing and thinking—all related to the mental faculty of an individual or a person. Any creative outcome of a person using his/her mental faculty is considered as an "Intellectual property", which is often intangible in nature. More specifically the term intellectual property refers to creative ideas resulting in patents, copyrights, trademarks, trade secrets, geographical indication and industrial design, etc. In fact the modern usage of the term intellectual property goes back to the later part of the eighteenth century with the founding of the Swiss Federal Office for Intellectual Property in Berne.

Increasingly the world community is recognizing the importance of intellectual property (IP) for the economic development of their countries and started framing their own IP laws. These are meant primarily to protect the interest of owners of such intellectual property and their nation. Such laws are now called Intellectual Property Rights. Motivation for such measures are i) creation of the mind is becoming more valuable in the information era, and ii) modern information and communication technology (ICT) renders easy piracy and copying of such information.

Though IP laws are country-specific, at the international level they are governed by organizations such as World Intellectual Property Organization (WIPO) and World Trade Organization (WTO). They exert pressure on their member countries to enact IP laws in harmony with the agreed provisions in the conventions. All these laws provide,

i. exclusive rights by allowing the owners of intellectual property to get benefit from the property they have created,

ii. financial incentive for the creation of an investment in intellectual property, and

iii. meeting associated research and development costs.

A joint research project of WIPO and the United Nations University analysed the impact of IP systems on six Asian countries and found a positive correlation between strengthening of the IP system and economic growth.

While more and more countries are introducing intellectual property laws, there appears some criticism also. Certain significant criticism include that it promotes monopoly and inhibits the flow of innovations to poorer nations. There is a tendency to expand intellectual property both in duration and scope especially in copyright protection. Despite all these, the intellectual property right statutes have come into existence in many countries and more and more provisions are being added in conformity with international agreements.

As a research scholar has potential to come out with inventions arising out of research work and in later life also, he/she should have knowledge of intellectual property laws. This chapter is only an introduction covering aspects like copyright, patent, etc. which are of much use to scholars working in the field of science and engineering. Moreover the field of IPR heavily treads in legal discipline.

18.2   SOME PERSPECTIVES OF IPR

18.2.1   Significance of Key Terms

To enable readers understand and appreciate subsequent discussions significance of certain key terms of IPR are given below.

Copyright   A copyright means the exclusive right to perform or authorize the performance of acts in respect of a work or any substantial part thereof. The subject matter of copyright, the thing protected, is called a 'work'. Copyright is an intangible property.

Patent   A patent is granted by the state to the creator of an invention thereby giving exclusive right to make, use and sell that invention for a limited period of time provided a maintenance fee is paid annually. The invention for which a patent is granted must be a product or process possessing novelty, creativity and utility.

Invention   Invention refers to the creative act of an individual or a group whereby a new and original idea is conceived which results in a product, process, etc. IPR covers invention.

Innovation   Innovation is the process by which an invention or an idea is brought into successful practice and is used by the economy. Innovation is not generally covered under IPR.

Trademark   A trademark is any name, word, symbol or a device that is used by a company to identify its goods or services and distinguish them from those made or sold by others.

Trade secret   A trade secret is any formula, pattern, device or compilation of information which is used in a business to create an opportunity over competitors who do not have this information.

Geographical indication (GI)   A geographical indication (GI) is a name or sign used on certain products that corresponds to a specific geographical location or origin like a town or a region or a country. The GI is used to serve as a certification to indicate certain specific qualities for products which enjoy reputation due to their geographical location.

Industrial designs   Industrial designs are also known as Design patents. It is a different type of patent and it covers ornamental aspects of a product including its shape, configuration, surface decoration, etc.

Industrial conventions/treaties   They are collaborative structures created by countries belonging to a region or globally to arrive at common procedures and steps for adoption by member countries. This is essentially to bring uniformity in laws regarding IPR to promote trade, research and development, etc.

18.2.2   Perspectives in Indian Context

Intellectual property right laws in India had their beginning in 1957. Since then the Government of India has taken several measures to enact new legislations, amendment of old legislations, etc. Further, India is a signatory to several conventions and also actively participates in these conventions and conferences relating to IPR. It has established statutory, administrative and judicial framework to safeguard intellectual property rights relating to copyrights, patent, trademark and industrial design.

Along with major economic reform measures and recognizing globalization imperatives, India took several initiatives to strengthen further the protection of intellectual property rights. These measures are intended not only to meet its own interests but also to comply with the provisions of international convention provision of which it is a member. Important Indian statutes available to protect IPR include,

i.   The Copyright (Amendment) Act, 1999, passed in December 1999 amending the old Copyright Act of 1957.

ii.   The Patents (Amendment) Act, 1999, enacted in March 1999 amending the Patents Act of 1970.

iii.   The Trade Marks Bill, 1999, which repeals and replaces the Trade and Merchandise Marks Act 1958 passed in December 1999.

iv.   The Patents (Second Amendment) Bill, 1999, to further amend the Patents Act 1970 and make it TRIPS complaint in December 1999.

v.   The Industrial Designs Bill, 1999, which replaced the Designs Act, 1911, and Plant Variety and Farmers Right Protection Act passed in 2001.

vi.   A sui generis legislation for the protection of Geographical Indication of Goods (Registration and Protection) Bill, 1999 and passed in December 1999.

vii.   The Design Act, 2000.

Not withstanding legislative measures, the Government of India introduced several measures towards modernization of patent information services and trademark registry with the assistance from WIPO/UNDP. Other measures taken up by the government include strengthening of administrative structure, additional infrastructure and modernization by way of computerization and re-engineering work practices. To create awareness about Intellectual Property Rights among public, the Central Government took several steps such as publishing of handbooks, organizing and sponsoring seminars and workshops, and organizing training programmes. Besides, the Government is also tightening its enforcement measures. It is reported that due to strict enforcement measures adopted by the government, there has been a declining trend in cases relating to violation of provision of Copyright Act.

18.2.3   The Global Scenario

Although many of the legal principles governing intellectual property have evolved over centuries, only in the twentieth century did some major developments take place. In USA the constitution empowers the congress to promote the programme of science and useful art. In UK some semblance of Copyright law and patent law existed in the seventeenth century itself though not exactly by the modern names. The Swiss had a federal office for intellectual property way back in 1888 itself. As far as Intellectual property laws are concerned, they are country-specific, i.e., each country enacts IPR laws according to their needs. However at the international level there is a trend towards harmonization of these laws in a collaborative way by creating conventions and treaties.

Both the Paris Convention of 1883 and the Berne Convention of 1886 got merged in 1893. This was originally located at Berne and subsequently relocated at Genova in 1960. At the instance of the United Nations this organization was succeeded in 1967 by a treaty and the World Intellectual Property Organization (WIPO) came into existence as an agency of UNO.

The WIPO Convention lays down the following list of activities or work which are covered under IPR.

- Industrial designs
- Scientific documents
- Protection against unfair competitions
- Literary, artistic and scientific works
- Invention in the field of human endeavour
- Performance of performing artists, phonograms and broadcasts.
- Trademarks, service marks and commercial names and designation
- All other rights resulting from intellectual activity in the industrial, scientific, literary or artistic fields.

The Trade Related Intellectual Property Rights (TRIPS) agreement has been largely successful in promoting a forum for countries to agree on an aligned set of patent laws. In fact conformity with the TRIPS agreement is a requirement for admission to the World Trade Organization (WTO) and many nations started giving importance to this. This has compelled many developing nations, which historically enacted laws for their development only, to look beyond and fall in line with global practices.

Major conventions and treaties in the global category are,

i.   Berne Convention for Copyright (1856)

ii.   Paris Convention for the Protection of Industrial Property (1967)

iii. Strasbourg Convention on International Patent Classification (1971)

iv.   Convention on Biological Diversity (CBD) (1992)

v.   General Agreement on Trade and Tariff (GATT) (1947)

vi.   Patent Cooperation Treaty (PCT) (1998)

India has been signatory in all conventions and treaties mentioned above. It takes active participation in all the conventions and conferences not only to protect their IPR issues but also that of developing countries in general. As India is a major contributor for computer software development, a lot of importance is given to protect IPR related to this field.

18.3 COPYRIGHT LAWS

18.3.1 Copyright Materials

Copyright is an intangible property. The subject matter of copyright, the thing protected is called a "work". Copyright exists throughout India in a) original literary, dramatic, musical and artistic works b) cinematograph film and c) sound recording. Section 14 of the Indian Copyright Act of 1957, subsequently amended defines that a "Copyright means exclusive right to do or authorize the doing of acts in respect of a work or any substantial part thereof". The original Act of 1957 was amended in 1983, 1984, 1992, 1994 and 1999. No copyright exists in any work except as provided in the Act. In this section, the discussion is restricted to only literary and computer-related aspects.

With regard to literary, dramatic or musical work the owner has exclusive right to reproduce work in any material form including its storage in any medium by electronic means to issue copies to the public, to perform or communicate the work to the public, and to translate or make any adaptation of the work. Literary work includes tables and compilation and therefore a dissertation is prima facie a literary work.

Literary work includes computer programs and compilation including computer databases as per sec.2(o), and sec.2(ffc) further elaborates on this. As far as computer programs are concerned the owner has exclusive right to sell or give on hire or offer for sale or hire any copy of the program. Amendments made in 1999 have further liberalized and added the right to give on commercial rental any copy of the program.

In order to harmonize with international practices, and in compliance with India's obligations under TRIPS, major legislative changes were brought in IPR. The Copyright Act was also amended. The member countries are required to comply with articles 1 to 21 of the Berne Convention. Under the literary work were included computer program tables and compilations including computer databases. Computer program means a set of instructions expressed in words, codes, schemes or in any form including a machine-readable medium, capable of causing a computer to perform a particular task or achieve a particular result.

The computer program being a literary work, should be original for being copyrighted. For this the author should spend sufficient labour, skill, judgment and capital to impart the work some quality or character to distinguish it from the original raw material. After the amendment to Copyright Act in 1999, India has complied with all the requirements of the Article in TRIPS.

18.3.2   Ownership and Assignments

The Copyright Act confers upon the author the economic rights to enable exploitation of the work by himself/herself or by licensing it for royalty which brings economic benefits to the author. Thus the objective is to encourage authors to do original work by giving them exclusive right for a limited period of time. It is a negative right to prevent others from copying their work.

In India, the copyright is awarded for the life of the author plus 60 years from the death of the author. For being the owner of copyright, registration is not compulsory. The author of a work becomes its owner thereof the moment the work is created. Registration is not necessary either for the subsistence or for the enforcement of copyright. As per section 48 of the Act, the Register of copyright is the prima facie evidence of particulars entered therein. Therefore, it is advisable to register the copyright in the work.

The ownership of copyright in a work is independent of the ownership to the physical material in which the work is fixed. For example, when you buy a book you become the owner of the book, but the author of the book is the owner of copyright unless he/she assigned the copyright to a publishing company. While the author of any work is the first owner of copyright, however when literary work is created by the author during the course of employment under a contract for service, the employer is the owner of copyright. A teacher who writes a book during his/her employment is the owner, since he/she is employed to teach and not to write.

A copyright can be assigned, but however it should be in writing and signed by the assignor or by a duly authorized agent (Sec.19). Copyright is a multiple right consisting of a bundle of different rights in the same work, which can be assigned or licensed either as whole to one party or separately to different parties.

18.3.3   Securing of Copyrights

Although a copyright need not be registered and it exists from the moment a work is created, the registration is an evidence to support his/her claim as a copyright holder. The following procedure has to be followed to secure copyright registration.

i.   Filling of application in the proper format to the Registrar of Copyrights in triplicate. In the case of an artistic work that can be used in some other product such as logo, emblem, mascot, trademark, etc., a statement has also to be included along with the application.

ii.   Separate application must be filed for each work along with necessary registration fees fixed by the Government depending on the nature of work.

iii.   The applicant for copyright has to provide copies of copyright application to all persons like co-authors, publishers and others interested in the copyright of the work.

iv.   Mandatory period allowed for receipt of objections from interested parties is 30 days from the date of filing of an application.

v.   If no objection is received from anybody, the Registrar of Copyrights scrutinizes the application and then enters the details given in the application in the Register of Copyrights kept for the purpose.

vi.   In case any objections are received and the Registrar is not satisfied with the correctness of application, he/she may conduct an enquiry and/or call for explanation from the applicant. The Registrar after thorough satisfaction, makes an entry in the register kept for that purpose.

vii.   A copy of entries made in the Copyright register is sent to all parties concerned as evidence of Copyrights granted.

18.3.4   Rights and Infringements

The Act recognizes several moral rights of the author, which exist even after assignment of the work. These rights include claim of authorship of the work, and to restrain and claim damages in respect of any distortion, mutilation, modification or other act in relation to the work which is done before the expiry of the term of copyright.

In the case of published work the copyright will subsist in India if the work is published in India or outside India and the author at the time of publication (if alive at that date) or at the time of his/her death is a citizen of India. In the case of unpublished work the copyright subsists in India, if the author at the time of making of the work was a citizen or domicile of India.

If any of the acts specified in section 14 relating to the work is carried out by a person other than the owner or without license from the owner or competent authority under the law, it constitutes infringement of copyright as per Sec 51 of the Act. The type of infringement will depend upon the nature of the work.

Certain acts will not be construed as infringements as per the law. Permitted acts include fair dealing, educational purpose, literary and archive use, use associated with public administration and certain acts carried out by lawful users of computer programs. The lawful users can take back-up copies de-compile the program in limited circumstances, copy and adopt programs in ways consistent with lawful use.

For violation of Copyrights, both civil and criminal remedies are provided in the Act. The remedy may be an injunction, claiming of damages or share of profits, etc. Courts are vested with powers to punish copyright violators for criminal offences. The police are empowered to confiscate without warrant pirated copies of the work including equipment used for the purpose.

India has one of the most modern Copyright protection laws in the world. The 1999 Amendment (49 of 1999) not only made it fully compatible with the provisions of TRIPS agreement, but also brought under Indian Copyright protection additional items such as satellite broadcasting, computer software and digital technology. The other important developments were the issuance of International Copyright order 1999 extending the provisions of Copyright Act to the nationals of all World Trade Organizations (WTO). The Copyright Act prescribes mandatory punishments for piracy of copyrighted matter commensurate with the gravity of the offence in order to deter infringement in compliance with TRIPS agreement.

18.4 PATENT RIGHTS

18.4.1 Patent Law Concepts

The word "patent" originates from the Latin word patare, which means "to lay open", i.e., to make available for public inspection. Now it is recognized that patent is a legal grant by a government to a person for his/her invention. As per Indian Patent Act, "invention" means a new product or process involving an inventive step and capable of industrial application. Governments establish patent systems and grant patents to encourage innovation, technical development, social recognition and ultimately economic benefits to the holder. The award is in the form of monopolistic rights for his/her labour and efforts. A patent holder has the rights to prevent others from making, using or selling the patented invention for a limited period of time.

Two basic issues are important in this. The first and most fundamental concept to know about patents is that patents do not accord the right to the holder the right to practice the invention claimed in the patent, but only exclude others from practising the invention. However, the patentee may only practise his/her invention as long as it, or any part of it, is not covered in a valid patent of others. The second basic concept to keep in mind is that a patent is effective only in the country where the patent was granted. Since governments are granting patents and all countries are different, naturally the patent laws will be different in every country around the world.

Countries across the world have realized and are also becoming partners of cooperative treaties, which allow inventors to file patents easily. Not only does it provide additional revenue by way of patent fees, but also provides incentive for business to invest in these countries. Such cooperative treaties have harmonized some of the basic procedures and requirements for obtaining patents from different countries. Notwithstanding this, substantial differences still remain regarding patent law.

The Indian Patent Law came into existence by way of the Indian Patent Act 1970 (Act 39 of 1970). Subsequently several amendments have been made and the notable one was in 1999. This amendment called the Patents (Amendment) Act 1999 was made to ensure harmony with international practices and was in compliance with India's obligation under Trade Related Intellectual Property Rights (TRIPS).

18.4.2 Items not Patentable

According to Indian Patent Act, the patentable inventions comprise new product or processes involving an inventive step and capable of industrial application. Instead of giving an exhaustive list of items which are patentable, the act clearly laid down items which cannot be considered as inventions and obviously not qualify for the grant of patent. The list of items include,

- Any invention or commercial exploitation or intended use which would be contrary to law or morality or injurious to health.
- Mere discovery of scientific principle or the formulation of an abstract theory.
- Discovery of any new form of known substance which does not result in the enhancement of the known efficacy of that substance or the mere discovery of any new property or new use for a known substance or the mere use of a known process, machine or apparatus unless such known process results in new product or employs at least one reactant.
- A substance obtained by mere admixture resulting only in the aggregation of the properties of the component thereof or a process for producing such substance.
- Mere arrangement or duplication of known devices, each functioning independently of one another in a known way.
- Traditional method of agriculture or horticulture.
- Mathematical or business method or a computer program per se or algorithm.

- A literary, dramatic, musical or artistic work or any other aesthetic creation whichever including cinematographic works and T.V. production.

- Topography of integrated circuits.

- Any process for the medicinal, surgical, curative prophylactic or other treatment of human beings or any process for the similar treatment of animals or plants.

- Any invention relating to atomic energy falling within the sub-section (1) of Section 20 of the Atomic Energy Act 1962 (Act 33 of 1962).

18.4.3   Securing of Patent Rights

To secure patent right in India, an inventor or his/her agent has to follow the procedures given below:

i.   Any person who is a true and first inventor, an assignee or the legal representative of a deceased inventor can make an application to the Controller General of Patents, Designs and Trademarks enclosing specifications, detailed drawings, etc.

ii.   The date on which the patent application either with or without provisional specification or complete specification is filed at the patent office is called date of priority.

iii.   The Controller General refers it to appropriate examiners for scrutiny and submission of reports.

iv.   If there are objections/clarifications needed based on the examiners report, the applicant is required to file necessary explanation within a time frame. In case no reply is received, the patent application is abandoned.

v.   Once the Controller after examination accepts the specification and issues a notice to the applicant and also advertises in the official gazetteer the fact that specification accepted, the drawings and specifications are open for public comments.

vi.   Four months time is given to file any comments by opponents and others. An opportunity is also given to them for hearing by the Controller.

vii.   Thereafter if everything is in order the Controller General grants the patent right to the applicant.

viii.   The period is usually 20 years from the date of patent provided the patentee pays maintenance charges annually. The date of patent is the date of application filed by the applicant.

ix.   An Indian resident cannot apply for patent for an invention outside the country without prior permission. He/she should have applied first to Indian authorities at least six weeks before making application to outside.

x.   While granting patent right, the authorities can impose conditions with regard to import and use by the Government, limited use of patented item by others for experiment or research and instructions to pupils, etc.

18.4.4 Rights of a Patentee

The Patent Act confers exclusive rights to the patentee to prevent third parties from making, using, offering for sale, selling or importing for those purposes that the product provides in India. This is applicable for both product and process. The grantee or the proprietor has the power to assign, grant license under or otherwise deal with the patent and to give effectual receipts for any consideration for any such assignment, license and dealing.

The authority can compulsorily give license any time after three years of expiry from the date of sealing (grant) of a patent to any other person applied for grant of compulsory license on the ground that reasonable requests of public with respect to the patent have not been satisfied, patented invention not available to the public at a reasonably affordable price and the patented invention not worked in India.

Certain general guiding principles have been enunciated to check whether the patented invention works or not. They are as follows:

- Whether the invention for which patent granted worked in India on a commercial scale without delay.
- Patents are not granted merely to enable patentees to enjoy monopoly for the importation of a patented article.
- Whether the protection and enforcement of patent right contributes to the promotion of technological innovation and to the transfer and dissemination of technology.
- That patents granted do not impede protection of public health and nutrition.
- That they do not in any way prohibit the Central Government in taking measures to protect public health.

The Act also provides for revocation of patent rights granted on the ground that the patented invention has not worked in India and that it has not satisfied the reasonable requirements of the public and that the patented invention is not available to the public at a reasonable price.

18.4.5 Securing of Global IPR

Normally grant of patent is accorded by the government of a country for an applicant and is valid only in that country. However if the researcher wants to get patent right for the same invention from other countries he/she should first apply in their own country and only then to other countries selected. There are international treaties, which enable global acquisition of intellectual protection easier and more uniform. In general these treaties are used to make the filing of patent application easier for the applicant. The actual filings are generally handled by the patent agent.

The Paris Convention for the Protection of Industrial Property, commonly called the "Paris Convention" is a multi-lateral treaty originated in 1883 and has been revised several times. Over 100 countries have become signatories to this body, which is administered by the World Intellectual Property Organization (WIPO) located in Geneva. India is a signatory to this convention. This treaty allows an applicant a year's grace period after filing an application in a signatory country to file applications in other signatory countries while still retaining the original filing date or priority date. Patent application, design patent application and trademark application are all covered by this convention.

Another important treaty known as the Patent Cooperation Treaty (PCT), which came into existence in 1970 has become a major method by which patent applications are filed internationally. The PCT is also administered by WIPO. India is a member country of this treaty. The prime advantage of the PCT is that it provides a convenient mechanism for filing patent applications for the same invention in many different countries at one time. As per this an applicant can file one set of papers in any one of several receiving offices located worldwide in the applicant's native language in a signatory country.

Normally an applicant does not use both Paris Convention and PCT for filing application. The Paris Convention is used to establish priority date for the application, while the PCT is the actual filing mechanism for most of the global applications. In this case, the inventor first files his/her application in the home country and then the PCT application specifying the countries in which he/she wants patent for IP within a year of the filing date. This procedure enables the home country to start examining the application and the inventor has some additional time to understand the value of the invention before paying fees for filing the PCT application.

The PCT provides several benefits for the applicant to get global IPR. Important benefits are,

- A single application is filed for the purpose of global patentability search report and also for claiming the priority date in all the specified countries.

- Application can be in the native language first, although translation may be required later.

- Opportunity to defer the payment of filing, translation and other fees for up to 30 months after the priority date of application.

- An application filed using PCT has an international search facility provided by one of the international search offices.

- An applicant can initially file for IP in almost all countries of the world for one fee, and reduce the number of countries at a later date without incurring additional fees.

There are several other conventions/treaties in the world similar to ICT like the European Patent Convention for European countries, the African Intellectual Property Organization (OPAI) with many African countries as members and Eurasian Patent

Organization (EAPO) mainly comprising countries which were part of the USSR earlier. All these regional conventions/treaties provide regional collaboration among their member countries.

18.4.6 Patent Search

Inventors, consultants, companies and R&D organizations resort to patent search to meet their varied needs. Some of the major purposes for which such searches made are,

- Bibliographic searches to find out what is covered by a specific patent granted to a particular patent owner.

- Information search to know the patent publications that already exist in the field of interest.

- Search by patent offices to examine a filed application based on similar standards pertaining to the utility, novelty and non-obvious requirements.

- Equivalent patent search to identify family members of a specific patent document.

- To assess whether the proposed product or process is infringing any of the aspects of a valid patent already granted.

- For long-term strategic decision on R&D, companies make searches for a period to trace all patents related to a selected technology or product/process.

It may be noted from the above that whatever be the objective, the required information has to be culled out from the computerized database of all patents granted by the respective countries and maintained by the authorities concerned.

Apart from individual inventors and companies, there are patent attorneys and patent agents who provide specialized patent search services, undertake filing of patent applications and pursue the matter till the patent is granted.

To conduct a patentability search, lot of information needs to be provided. Major information requirements include illustrations or photographs of patent with details of how it is made, any known similar designs, brief summary of shortcomings, advantages, possible applications and possible variants, etc. In addition to online facilities, many countries bring out periodical publications through their patent office, official Gazette, annual index, etc. In India an annual publication is brought out by the Controller General of Patent and Trade Mark in addition to maintenance of computerized database with accessibility countrywide.

In India, the Controller General of Patent and Trade Mark is the empowered authority for the issue of Patent rights. Their website is http://www.patentoffice.nic. in/iprs1/patentsearch.htm. A patent search system called Indian Patent Information Retrieval System (IPIRS), has been implemented. Using this online facility, one can get information about i) application status both after publication till the stage of grant and information after grant, ii) information about patents granted, and

iii) published applications. The system provides for a single parameter search both in quick search made and advanced search made. The website provides adequate guidelines to make the search easier and an interested reader may refer to the website.

18.5 OTHER IPRs

18.5.1 Trademarks

Trademarks are names or symbols used by manufacturers to identify their products differentiating from their competitor's products. Manufacturers not only use this to maintain their reputation but also to prevent others from copying their product and offer a sub-standard quality product using their name. Normally any word or symbol can be used as a trademark by the manufacturer at the same time ensuring that there is no conflict with another trademark. A trademark should meet the guidelines prescribed by the trademark granting authority of the country. Like trademarks, service marks are also used to protect the services offered by the service providers.

A trademark granted by the appropriate authority gives protection to the owner of the mark the exclusive right to use it to identify the goods manufactured or service provided. The owner can also license the trademark to his/her customers for use with their products and also prevent them from using it. In India the Controller General of Patent and Trade Mark is the authority to grant trademark rights. The duration of the exclusive right to keep a trademark is ten years. However, one can also keep it permanently by renewing it once in ten years.

To make it known to the third parties that word written in a particular manner is a trademark or service mark, abbreviations 'TM' and 'SM' are used as notations. These notations are normally used as superscripts like "Coca-Cola^{TM}" once an application is filed for that. After the grant of trademark right, the symbol "®" can be used instead of superscript 'TM'. The governing regulatory statute is the Indian Trademark Act, 1999.

18.5.2 Geographical Indication (GI)

It is common to see a product enjoy reputation because of its special qualities which is peculiar to a particular city, region or a country from where it is originated. For instance, Kancheepuram Silk sarees are world-famous because of the special method of weaving using silk threads at Kancheepuram. The word 'Kancheepuram' is an indication of the location of origin of this particular variety of silk sarees. Thus, a geographical indication is a name or sign used on certain products that corresponds to a specific location or origin. In fact use of GI serves as a certification, besides preserving the cultural traditions of the locality.

The Indian Geographical Indication Act, 2001, is the statute which regulates the registration and protection. Any association of persons, producers, organization or authority can apply for GI representing the interests of producers of the product. While filing application, they should indicate the origination of the product, the quality, reputation and other characteristics pertaining to that geographic origin, etc. The Registrar of Geographical indications is the authority empowered to grant registration.

The registration process requires filing of one application by the organization concerned. After initial scrutiny, the Registrar constitutes a consultative group of experts to ascertain the correctness of particulars mentioned in the application. The Registrar calls for explanation/additional details if needed and then issues a notification calling for objections in the geographical indication journal. After consideration of objections if any received, the Registrar registers the GI in Part-A of the register. The authorized users of GI have to register separately in Part-B with the Registrar.

Both WIPO sponsored international treaties and agreement on TRIPS offers protection at an international level to GI within the framework of the World Trade Organization. A registered GI is protected by several laws such as laws against unfair competition, consumer protection laws, laws for the protection of the certification marks, etc. The protection includes legal remedies which the owner of GI can seek, for any violation by others. Since a GI is considered a public property belonging to the producers of a product in a particular area, no assignment, licensing, pledge, etc. is permitted. As in the case of trademark, the validity of GI is also ten years and it is extendable for a further period of ten years.

18.5.3 Trade Secret

A trade secret is considered as a critical information and a know-how pertaining to a business undertaking, which they want to keep it out of public domain. It allows the person who holds it to make money on monopolistic basis because others are not aware of the information. A trade secret could be a formula, a production process, computer software, device, technique, etc. But, it is the responsibility of the owner of trade secret to take extra precaution to maintain the security of the trade secret. As this is maintained by the organization/owner they can hold it indefinitely.

While patents and copyrights require their holders to disclose their information as a part of the application process, the trade secret requires the holder to keep the information secret by themselves and is not required to be disclosed. The duration of the trade secret is much longer compared to patents or copyright and it can potentially last forever. Trade secrets are valid only as long as others do not discover and make it public. If the trade secret is made public by anybody in violation of the non-disclosure agreement, then the holder has the right to take legal recourse to sue and claim damages from the violator.

There are no subsisting legislations in India at present to protect trade secrets. Nevertheless, as a part of its obligation under WTO and also to comply with the conditions of TRIPS, Idia has agreed to protect undisclosed information.

Glossary

Abstracting service A service provided by libraries in the form of summary of the content of papers published in journals for the benefit of researchers and others.

Accuracy Refers to the agreement between a measurement and its true value.

Algorithm A sequential mathematical procedure or a set of rules or steps employed for calculation or solving a problem.

Analysis of variance (ANOVA) A statistical technique used to test the significance of some parameters involving multiple samples simultaneously by cross-classifying the variances and testing.

Appendices Supplementary information and data pertinent to the study which are not important but useful to appreciate certain points, are provided under appendices. This is done to improve the flow of presentation of the main body of the report.

Applied research One type of research concerned with finding a solution for an immediate problem faced by the society or an industrial/business organization.

Arithmetic mean (AM) Most common and an important measure of central tendency of an array of data, computed as the average of all data points.

Bibliography Term used to present the list of all works which a researcher has read and to which he/she is indebted for ideas or information in general.

Black-box model A mathematical model in which there is no prior information available.

Blocking design A type of experimental design that uses blocking techniques to remove the effect of background variables from the experimental error.

Buckingham π theorem A theorem used to describe a set meaningful of equations involving n variables into an equivalent set of equations involving (n − m) dimensionless variables, where m is the number of fundamental dimensions used.

Citation index An index based on the principle of existence of some meaningful relationship between the papers published by two or more authors. It is amenable for analysis and serves as a criterion for judging the quality of the paper.

Conclusions   Refers to the body of logical inferences drawn and final views formed arising out of findings from a research study.

Confidence interval   Also known as confidence limits, it is a statistically established interval in which a parameter lies with some specified probability.

Confounded relationship   When the dependent variable is likely to be influenced by extraneous variables, the relationship between dependent and independent variable is said to be confounded by the extraneous variable.

Copyright   An intangible property and exclusive right conferred to an author to do or authorize the doing of acts in respect of his/her work or any substantial part thereof.

Correlation   Indicates the degree of association between variables and is expressed as a coefficient, a number between -1 to $+1$. The -1 indicates perfect negative correlation, $+1$ indicates perfect positive correlation and 0 indicates no correlation.

Creativity   A natural and normal ability of a human being possessed in quantity as a part of genetic structure which is unique.

Cybernetics   Science of communications and automatic control system related to man–machine system.

Descriptive and diagnostic research   Descriptive research studies are those studies used to characterize a particular material, process, individual or a group, etc. The diagnostic studies are concerned with the frequency with which something occurs and its association with something else.

Design of experiments (DOE)   Also known as experimental design, it deals with design of any information gathering exercise where variation is present whether under the full control of experimenter or not.

Dimension analysis   An important form of symbolic modelling, used as an aid in modelling and experimentation when the functional relationship between design variables are not well-established or are not known.

Dissertation abstract   A compilation of the abstracts of doctoral theses of participating universities in USA and Canada, which is made available for use by interested parties and academia.

Editing   The process of critically examining the rough draft and employing self-criticism so as to strengthen the draft by analysing each paragraph, sentence structure, use of words, checking of spelling, punctuation, etc.

Empirical models   A set of mathematical models developed using a set of data obtained by empirical means such as experiment or observation, etc.

Error   Used synonymously with uncertainty, it is the difference between an observed or computed value and a true or theoretically correct value.

Error bars   Graphical representation of standard error of measurement at each value by superimposing it on the graph of mean value of a parameter of interest.

Ethics   Defined as a system or code of morals of a particular person, religion, group or profession and differs from one individual to another.

Evolutionary algorithms   Algorithms based on heuristics that use natural selection process to search for optimal solution in solving problems.

Experience survey   Method of survey conducted by the researcher with experts who have practical knowledge of the problems studied to get greater insights into the relationship between variables and fresh ideas related to the research problem.

Explanatory research   A type of research study with emphasis on discovery of concepts (ideas) and better insights.

Extraneous variable   Independent variables that are not related to the objective of the study but affect the dependent variable.

Factorial designs   An experimental design in which all the levels of each factor are combined with the levels of all other factors to conduct the experiment.

Factors   Refers to input or experimental variables that are supposed to be controlled by the experimenter while conducting experiments.

Footnotes   Notes used to validate a point/statement to supplement or amplify the materials included in the main body and invariably given at the bottom of the respective page.

Frequency distribution   A method of expressing voluminous data by arranging them into a number of equal class intervals and then determining the frequency of the observations falling in each class interval. Expressed in a tabular and/or graphical form (histogram).

Fundamental research   This type of research is concerned with some natural phenomena or some physical principles aimed at finding information having broad base of application.

Genetic algorithm (GA)   An evolutionary search optimization technique which works by mimicking the evolutionary principles and chromosomal processing in natural genetics.

Goodness of fit   A statistical approach to check whether the observed or empirical data fits an assumed theoretical distribution.

h-index   A measure used to indicate both the scientific productivity and the apparent impact of scientists. It is established on the basis of the number of citations that a scientist or author's paper received in other people's publications.

Impact factor   An index to assess the quality of a journal, and calculated by dividing the references cited in one year by the total number of papers published in the same journal over the previous two years.

Industry research   A type of research directed towards innovation, introduction and improvement of products and process related to an industry.

Intellectual honesty   The researchers' honesty in presenting proposals, data and in communicating their best understanding of their work both in writing and verbally.

Intellectual property   Refers to creative ideas resulting in patents, copyrights, trademarks, trade secrets, geographical indication, industrial design, etc.

Invention and innovation   Invention arises out of creative and original idea of someone whereas innovation refers to the process by which an invention or idea is translated into successful practice.

LabVIEW   An interactive software containing graphic modules called virtual instruments, used by researchers for varied type of modelling and analysis.

Level of significance   Indicated as a percentage (usually 1% or 5%) which reflects the risk a researcher is willing to take in rejecting a null hypothesis, when it is true.

Literature review   The examination of relevant literature by the scholar to review the critical points of current knowledge and methodologies adopted on a particular topic of interest to the scholar.

Literature survey   The survey of literature such as scholarly articles, books, dissertation, proceedings, etc. both online and offline so as to gather information on a topic of interest by a researcher to meet his/her varied needs.

Mathematical model   A type of model developed for a specific purpose by employing a set of variables and a set of equations that establish relationship between variables.

MATLAB   An interactive computational software used by researchers for analysis.

Model   An abstraction or idealization of an object, idea, system, real or hypothetical, in some form other than the entity itself. They are classified as dynamic or static, deterministic or probabilistic and iconic or analog.

Model validation   A process of quantifying uncertainties of a model by comparing model predictions with the real-world data to reinforce confidence in the model developed.

Optimization   The act of obtaining the best results for the objective of interest taking into account all constraints and other conditions.

Orthogonal array   A kind of special matrices that permits the experimenter to study the effect of several parameters efficiently.

Outlier   A wild or questionable data point which does not follow the pattern of other data points of results in a study or an experiment.

Parameter In statistics, this term is used to represent items of interest such as mean, standard deviation, etc. when they refer to population or universe, e.g., population mean (μ), population standard deviation (σ)].

Particle swarm optimization (PSO) A computational search technique employed to derive optimal solution to a problem which works on the principle of collective behaviour of biological species like birds, bees, etc.

Patent A right conferred by the state to the creator of an invention thereby he/she can make, use and sell that invention to another person or body.

Patent search Searching the database in the website maintained by the Controller General of Patents and Trademarks by inventors, consultants, companies and R&D organization for their varied needs.

Plagiarism The act of an author taking credit or attempting to take credit for the work of another.

Population Refers to all items of interest such as individuals or their attributes or results of operation which may be finite or infinite and subject to investigation. It is also referred to as 'universe'.

Precision Indicates the agreement between repeated measurement of the same parameter, thus a measure of repeatability.

Probability Defines the likelihood of an occurrence of an event and is expressed as a number between 0 and 1.

Probability density function (PDF) Referred to as density of a continuous random variable expressed as a function describing the occurrence of the random variable at a given point.

Probability mass function (PMF) In statistics it represents the probability that a discrete random variable is exactly equal to some value and is the primary method of defining a discrete probability distribution.

Problem definition In research, this comprises the topic identified, objective and hypothesis proposed, data to be gathered, experiment and technique of analysis to be used and the boundary of the problem studied.

Random error A type of error that fluctuates from one measurement to the other in a random fashion and is distributed about some mean value.

Randomization A concept used in experimental design in which the various specimens to be considered for the study are selected in a random fashion to minimize variation due to extraneous factors.

References Term used to indicate the specific location of an idea or a piece of information in the original source which a researcher has used or cited.

Referencing Indicates the various methods of citing the work of others published in journals, books and other sources to support a researcher's point or to develop further on the previous findings. Several styles of referencing are there.

Regression analysis A statistical technique used to establish an appropriate mathematical relationship between the dependent variable and independent variable(s) for prediction purposes.

Replication Refers to the repetition of experiment for the desired number of times in the context of design of experiment to improve reliability of results.

Research A systematic and critical investigation to establish new facts and conclusions thereby widening the frontiers of knowledge.

Research design 1. An action plan dealing with issues such as manner of gathering data, sources and conditions, tools and techniques to be used and methods of analysis for the research study taken up, to ensure credibility and smooth sailing. 2. A plan for the collection of relevant data/evidence with minimum effort and time, which depends upon the nature of the research work.

Research ethics Application of fundamental ethical principles to a variety of topics related to scientific research.

Research hypothesis A tentative statement about the outcome of a study. It could be statement relating to an independent variable with the dependent variable which can be tested statistically.

Research methodology It differs from research method in that it not only includes research methods but also the logic behind the methods a researcher employs in the research study.

Research methods The methods that researchers use in performing research operations like relating the available data with the unknown aspects of the problem to derive solutions.

Research process The series of actions or steps necessary to carry out a research study properly and the sequence involved in it.

Sample A small portion or part selected from the population for the purpose of critical study/experimentation. The process of selection is called sampling.

Sample design A plan that a researcher draws for selecting samples from the population, as the study of entire population is neither possible nor economical.

Sampling distribution Represents the distribution of sample statistics such as mean ($\bar{x}$), standard deviation (s), when several samples are drawn from the population.

Scatter plot matrix A sequence of graphs of dependent variables versus each of the independent variables arranged in the form of a matrix for drawing useful conclusions.

Scientific method Refers to all research methods and techniques in general though there are variations between one science and another. It encourages a rigorous, impersonal mode of procedure dictated by the demands of logic and objective procedure.

Scientific misconduct Intentional distortion of a research process by fabrication of data, text, hypothesis and/or methods from another researcher's manuscript or distortion of research process in other ways.

Search engine A kind of software that enables the user to search for information on worldwide web (www) through internet.

Similitude Concept used in testing of engineering models. A model is said to have similitude with real object/system only if the two share similarities such as geometric, kinematic and dynamic.

Standard deviation Another measure of dispersion obtained as the positive square root of variance.

Standard error The standard deviation of a sampling distribution used to check whether the observed distribution of an experimental data differs significantly from the expected distribution.

Statistic This term is used to represent items of interest such as mean, standard deviation, etc. when they refer to the sample [e.g., sample mean ($\bar{x}$), sample standard deviation (s)].

Statistical probability Known as empirical probability of an event, it is estimated as the relative frequency of occurrence of an event when the number of observations is very large.

Statistics A science of collection, correlation, summarization, presentation and analysis of data as well as drawing of inferences and making reasonable decisions on the basis of such analysis.

Synopsis A concise outline survey or summary of a doctoral work of a scholar to be written and submitted to the university for approval before a thesis is prepared.

Systemic error Also denoted as bias error, which tends to shift all measurements in a systematic way so that their mean value is displaced in a particular way.

System simulation A process of designing a model of a real or a hypothetical system and conducting experiments with this model to understand the behaviour of the system or evaluating strategies for the operation of the system.

Theoretical models Represents models that are developed using physical laws, material properties, equilibrium conditions, etc. to link the input variable(s) with the response variable depending on the context.

TRIPS Trade Related Intellectual Property Rights. An international agreement which the signatories should comply with for promotion of harmony of laws of their country related to intellectual property. It is a prerequisite for a country to be a member of World Trade Organization.

Variance A measure of dispersion of data, calculated as the sum of squared deviation of all data from the mean and dividing the sum by the number of data points.

Verification Refers to ensuring that a model, especially simulation model, behaves as the experimenter intends.

Vertical science   Any discipline of science in which results are built upon results which in turn are built upon results. For example Physics is a vertical science.

White-box model   A mathematical model in which all necessary information is available.

Working hypothesis   A tentative proposition made to establish and test its logical or empirical consequences, which is necessary for the researcher to draw the boundary of research work.

World Wide Web (www)   An enormous collection of colourful on-screen documents that are linked to each other by highlighted words called hypertext and supported by friendly graphical interface. They are assessed through internet using search engines.

References

Abdul Rahim, R. (1996). Thesis Writing: A Manual for Researchers. New Age International (P) Ltd., New Delhi.

Abubacker, K.M., Sathish Singh, D., Mannar Jawahar, P. and Ganesan, R. (2004). "Kinematic optimization of 18 speed gearbox using genetic algorithm." International Conference and Exhibition on Total Engineering Analysis and Manufacturing Technologies, Bangalore, 2–4th, November.

Abubacker, K.M., Sathish Singh, D., Mannar Jawahar, P. and Ganesan, R. (2005). "Design optimization of spur gears using genetic algorithm." Proceedings of National Conference on Competitive Manufacturing Technology and Management for Global Marketing, Chennai, 7–8th, January.

Ahmed Helmy. "A personal note on how to start research in computer networks: seven steps on road to success." Available at http://www.cise.ufl.edu/~helmy.

Brian Tracy. (2005). Goals! Tata McGraw-Hill Publishing Co. Ltd., New Delhi.

Buckheit, J.B. and Donoho, D.L. (1995). Wavelab and Reproducible Research. Department of Statistics, Standford University, Technical Report 474.

Cochran, W.G. (1963). Sampling Techniques, 2nd edn. John Wiley and Sons, New York.

Dan Jones. (1998). Technical Writing Style. Pearson Education Company, Massachusetts.

Dickey John, W. (1975). Metropolitan Transportation Planning. Scripta Book Company, Washington D.C.

Dieter, G.E. (2000). Engineering Design: A Materials and Processing Approach, 3rd edn. McGraw-Hill, Book Co., Singapore.

Doebelin, E.O. (1995). Engineering Experimentation: Planning, Execution, Reporting, 3rd edn. McGraw-Hill, USA.

Douglas, J.F., Gasiorek, J.M. and Swaffield, J.A. (1995). Fluid Mechanics, 3rd edn., Addison-Wesley Longman Ltd., U.K.

Fishman, G.S. and Kiviat, P.J. (1967). "The analysis of simulation generated time series." Management Science. Vol 13, No.7. pp. 525–557.

Frank, M. White. (2008). Fluid Mechanics. Tata McGraw-Hill Publishing Co. Ltd., New Delhi.

Ganesan, R. (1980). Corporate Simulation Study of Urban Bus System, unpublished Ph.D. thesis, Indian Institute of Technology, Madras.

Gay, L.R. (1996). Educational Research: Competencies for Analysis and Applications, 5th edn., Prentice Hall, Inc., New Jersey.

Government of India. (2010). Intellectual Property Laws: Acts, Rules and Regulations, Universal Law Publishing Co. Pvt. Ltd., Delhi.

Gurumani, N. (2006). Research Methodology for Biological Sciences. MJP Publishers, Chennai.

Hamdy, A. Taha. (2001). Operations Research: An Introduction, 6th edn. Prentice Hall of India Pvt. Ltd., New Delhi.

Holbrook, A., Bourke, S., Lovert, T. and Dally, K. (2004). "Investigating Ph.D. thesis examination reports." International Journal of Education Research. Vol. 41, No.2, pp. 98–120.

Holman, J.P. (2007). Experimental Methods for Engineers. Tata McGraw-Hill Publishing Co., Ltd., New Delhi.

IITM. "Standard Format for Preparing the Synopsis," Department of Electrical Engineering, IITM, India.

Illustrated Oxford Dictionary. (2006). Dorling Kindersley Ltd., and Oxford University Press, Penguin Books, India.

Intellectual Property Rights; Available at http://www.indianindustry.com/intellectual property-rights/

Kalyanmoy Deb. (2001). Genetic Algorithms for optimization. IIT, Kanpur, KanGAL,Report No.2001002.

Kennedy, J. and Eberhart, R. (1995). "Particle Swarm Optimization", Proc. IEEE conference on Neural Networks (Perth, Australia), IEEE Service Centre, Piscataway, N.J., pp. 1942–1948.

Kline, S.J. and McCintock, F.A. (1953). "Describing uncertainties in single sample experiments." Mechanical Engineering, Vol. 75. pp. 3–8.

Knight, H. Jackson. (2007). Patent Strategy for Researchers and Research Managers, 2nd edn., John Wiley and Smith, New York.

Kothari, C.R. (2004). Research Methodology–Method and Techniques, 2nd edn., New Age International (P) Ltd., New Delhi.

Madhav, S. Phadke. (1989). Quality Engineering using Robust Design. Prentice Hall, Englewood Cliffs, New Jersey.

Moffat, R.J. (1988). Describing the Uncertainties in Experimental Results, Experimental Thermal and Fluid Science. Elsevier Science Publishing Co.Inc., New York. 1: 3–17.

Murugan, M. (2000). Detection of predetermined tool wear and tool wear characterization, unpublished Ph.D. thesis, IIT, Madras, August.

Nasreen Haider (2008). "Finite element analysis of the effect of cutting edge radius in orthogonal cutting", unpublished M.S. thesis, Anna University, May.

National Academy of Sciences. (2009). On Being a Scientist, 3rd edn., The National Academic Press, Washington. Available at: http://www.nap.edu/catalog.php?Record id =12192 Patent search; Available at http://www.patentoffice.nic.in/iprsi/patentsearch. htm

RajaPrabu, R. (2004). Insulation characteristics of silicone and EPDM polymeric blends for high voltage applications, unpublished Ph.D. thesis, Anna University.

Rajasekaran, S. and Vijayalakshmi Pai, G.A. (2003). Neural Networks, Fuzzy Logic, and Genetic Algorithms, Synthesis and Applications. Prentice Hall of India, New Delhi.

Ramadoss, A. and Wilson Aruni, A. (2009). Research and Writing: Across the Disciplines. MJP Publishers, Chennai.

Rao, S.S. (1996). Engineering Optimization–Theory and Practice, 3rd edn. New Age International (P), New Delhi.

Research Ethics, Available at http://en.wikipedia.org/wiki/Research_ethics

Richard, A. Johnson. (2002). Probability and Statistics for Engineers, 6th edn. Pearson Edition, Indian Reprint.

Sankaranarayanan, V. (2010). Personal Communication, Department of Information Technology, B.S. Abdur Rahman University, Chennai.

Shannon, R.E. (1995). System Simulation: the Art and Science. Prentice Hall, Inc. Englewood Cliffs, New Jersey.

Shi, Y.H. and Eberhart, R.C. (1999). Empirical study of particle swarm optimization, IEEE Conference, pp. 1945–50.

University of New South Wales, "How to write a Ph.D. thesis", Sydney, Australia, Available at http://www.scribd.com/doc/29592539/howtowriteaph.d.thesis

Von Horn, R. L. (1971). "Validation of simulation results." Management Science. Vol 17, No. 5., pp. 247–258.

Warrier, B.S. (2006). "The defense of your thesis." Article in the Hindu-Education Plus, 4th December.

Warrier, B.S. (2006). "The making of a research scholar." Article in the Hindu-Education Plus, 18th September.

Woods R.L. and Lawrence, K.L. (1997). Modelling and Simulation of Dynamic Systems. Prentice Hall International Inc., New Jersey.

Index